HISTORY OF ANIMALS
BY

Aristotle

Translated by Theodorus Gaza

The History of Animals

Book I

Part 1

OF the parts of animals some are simple: to wit, all such as divide into parts uniform with themselves, as flesh into flesh; others are composite, such as divide into parts not uniform with themselves, as, for instance, the hand does not divide into hands nor the face into faces.

And of such as these, some are called not parts merely, but limbs or members. Such are those parts that, while entire in themselves, have within themselves other diverse parts: as for instance, the head, foot, hand, the arm as a whole, the chest; for these are all in themselves entire parts, and there are other diverse parts belonging to them.

All those parts that do not subdivide into parts uniform with themselves are composed of parts that do so subdivide, for instance, hand is composed of flesh, sinews, and bones. Of animals, some resemble one another in all their parts, while others have parts wherein they differ. Sometimes the parts are identical in form or species, as, for instance, one man's nose or eye resembles another man's nose or eye, flesh flesh, and bone bone; and in like manner with a horse, and with all other animals which we reckon to be of one and the same species: for as the whole is to the whole, so each to each are the parts severally. In other cases the parts are identical, save only for a difference in the way of excess or defect, as is the case in such animals as are of one and the same genus. By 'genus' I mean, for instance, Bird or Fish, for each of these is subject to difference in respect of its genus, and there are many species of fishes and of birds.

Within the limits of genera, most of the parts as a rule exhibit differences through contrast of the property or accident, such as colour and shape, to which they are subject: in that some are more and some in a less degree the subject of the same property or accident; and also in the way of multitude or fewness, magnitude or parvitude, in short in the way of excess or defect. Thus in some the texture of the flesh is soft, in others firm; some have a long bill, others a short one; some have abundance of feathers, others have only a small quantity. It happens further that some have parts that others have not: for instance, some have spurs and others not, some have crests and others not; but as a general rule, most parts and those that go to make up the bulk of the body are either identical with one another, or differ from one another in the way of contrast and of excess and defect. For 'the more' and 'the less' may be represented as 'excess' or 'defect'.

Once again, we may have to do with animals whose parts are neither identical in form nor yet identical save for differences in the way of excess or defect: but they are the same only in the way of analogy, as, for instance, bone is only analogous to fish-bone, nail to hoof, hand to claw, and scale to feather; for what the feather is in a bird, the scale is in a fish.

The parts, then, which animals severally possess are diverse from, or identical with, one another in the fashion above described. And they are so furthermore in the way of local disposition: for many animals have identical organs that differ in position; for instance, some have teats in the breast, others close to the thighs.

Of the substances that are composed of parts uniform (or homogeneous) with themselves, some are soft and moist, others are dry and solid. The soft and moist are such either absolutely or so long as they are in their natural conditions, as, for instance, blood, serum, lard, suet, marrow, sperm, gall, milk in such as have it flesh and the like; and also, in a different way, the superfluities, as phlegm and the excretions of the belly and the bladder. The dry and solid are such as sinew, skin, vein, hair, bone, gristle, nail, horn (a term which as applied to the part involves an ambiguity, since the whole also by virtue of its form is designated horn), and such parts as present an analogy to these.

Animals differ from one another in their modes of subsistence, in their actions, in their habits, and in their parts. Concerning these differences we shall first speak in broad and general terms, and subsequently we shall treat of the same with close reference to each particular genus.

Differences are manifested in modes of subsistence, in habits, in actions performed. For instance, some animals live in water and others on land. And of those that live in water some do so in one way, and some in another: that is to say, some live and feed in the water, take in and emit water, and cannot live if deprived of water, as is the case with the great majority of fishes; others get their food and spend their days in the water, but do not take in water but air, nor do they bring forth in the water. Many of these creatures are furnished with feet, as the otter, the beaver, and the crocodile; some are furnished with wings, as the diver and the grebe; some are destitute of feet, as the water-snake. Some creatures get their living in the water and cannot exist outside it: but for all that do not take in either air or water, as, for instance, the sea-nettle and the oyster. And of creatures that live in the water some live in the sea, some in rivers, some in lakes, and some in marshes, as the frog and the newt.

Of animals that live on dry land some take in air and emit it, which phenomena are termed 'inhalation' and 'exhalation'; as, for instance, man and all such land animals as are furnished with lungs. Others, again, do not inhale air, yet live and find their sustenance on dry land; as, for instance, the wasp, the bee, and all other insects. And by 'insects' I mean such creatures as have nicks or notches on their bodies, either on their bellies or on both backs and bellies.

And of land animals many, as has been said, derive their subsistence from the water; but of creatures that live in and inhale water not a single one derives its subsistence from dry land.

Some animals at first live in water, and by and by change their shape and live out of water, as is the case with river worms, for out of these the gadfly develops.

Furthermore, some animals are stationary, and some are erratic. Stationary animals are found in water, but no such creature is found on dry land. In the water are many creatures that live in close adhesion to an external object, as is the case with several kinds of oyster. And, by the way, the sponge appears to be endowed with a certain sensibility: as a proof of which it is alleged that the difficulty in detaching it from its moorings is increased if the movement to detach it be not covertly applied.

Other creatures adhere at one time to an object and detach themselves from it at other times, as is the case with a species of the so-called sea-nettle; for some of these creatures seek their food in the night-time loose and unattached.

Many creatures are unattached but motionless, as is the case with oysters and the so-called holothuria. Some can swim, as, for instance, fishes, molluscs, and crustaceans, such as the crawfish. But some of these last move by walking, as the crab, for it is the nature of the creature, though it lives in water, to move by walking.

Of land animals some are furnished with wings, such as birds and bees, and these are so furnished in different ways one from another; others are furnished with feet. Of the animals that are furnished with feet some walk, some creep, and some wriggle. But no creature is able only to move by flying, as the fish is able only to swim, for the animals with leathern wings can walk; the bat has feet and the seal has imperfect feet.

Some birds have feet of little power, and are therefore called Apodes. This little bird is powerful on the wing; and, as a rule, birds that resemble it are weak-footed and strong winged, such as the swallow and the drepanis or (?) Alpine swift; for all these birds resemble one another in their habits and in their plumage, and may easily be mistaken one for another. (The apus is to be seen at all seasons, but the drepanis only after rainy weather in summer; for this is the time when it is seen and captured, though, as a general rule, it is a rare bird.)

Again, some animals move by walking on the ground as well as by swimming in water.

Furthermore, the following differences are manifest in their modes of living and in their actions. Some are gregarious, some are solitary, whether they be furnished with feet or wings or be fitted for a life in the water; and some partake of both characters, the solitary and the gregarious. And of the gregarious, some are disposed to combine for social purposes, others to live each for its own self.

Gregarious creatures are, among birds, such as the pigeon, the crane, and the swan; and, by the way, no bird furnished with crooked talons is gregarious. Of creatures that live in water many kinds of fishes are gregarious, such as the so-called migrants, the tunny, the pelamys, and the bonito.

Man, by the way, presents a mixture of the two characters, the gregarious and the solitary.

Social creatures are such as have some one common object in view; and this property is not common to all creatures that are gregarious. Such social creatures are man, the bee, the wasp, the ant, and the crane.

Again, of these social creatures some submit to a ruler, others are subject to no governance: as, for instance, the crane and the several sorts of bee submit to a ruler, whereas ants and numerous other creatures are every one his own master.

And again, both of gregarious and of solitary animals, some are attached to a fixed home and others are erratic or nomad.

Also, some are carnivorous, some graminivorous, some omnivorous: whilst some feed on a peculiar diet, as for instance the bees and the spiders, for the bee lives on honey and certain other sweets, and the spider lives by catching flies; and some creatures live on fish. Again, some creatures catch their food, others treasure it up; whereas others do not so.

Some creatures provide themselves with a dwelling, others go without one: of the former kind are the mole, the mouse, the ant, the bee; of the latter kind are many insects and quadrupeds. Further, in respect to locality of dwelling place, some creatures dwell underground, as the lizard and the snake; others live on the surface of the ground, as the horse and the dog. make to themselves holes, others do not

Some are nocturnal, as the owl and the bat; others live in the daylight.

Moreover, some creatures are tame and some are wild: some are at all times tame, as man and the mule; others are at all times savage, as the leopard and the wolf; and some creatures can be rapidly tamed, as the elephant.

Again, we may regard animals in another light. For, whenever a race of animals is found domesticated, the same is always to be found in a wild condition; as we find to be the case with horses, kine, swine, (men), sheep, goats, and dogs.

Further, some animals emit sound while others are mute, and some are endowed with voice: of these latter some have articulate speech, while others are inarticulate; some are given to continual chirping and twittering some are prone to silence; some are musical, and some unmusical; but all animals without exception exercise their power of singing or chattering chiefly in connexion with the intercourse of the sexes.

Again, some creatures live in the fields, as the cushat; some on the mountains, as the hoopoe; some frequent the abodes of men, as the pigeon.

Some, again, are peculiarly salacious, as the partridge, the barn-door cock and their congeners; others are inclined to chastity, as the whole tribe of crows, for birds of this kind indulge but rarely in sexual intercourse.

Of marine animals, again, some live in the open seas, some near the shore, some on rocks.

Furthermore, some are combative under offence; others are provident for defence. Of the former kind are such as act as aggressors upon others or retaliate when subjected to ill usage, and of the latter kind are such as merely have some means of guarding themselves against attack.

Animals also differ from one another in regard to character in the following respects. Some are good-tempered, sluggish, and little prone to ferocity, as the ox; others are quick tempered, ferocious and unteachable, as the wild boar; some are intelligent and timid, as the stag and the hare; others are mean and treacherous, as the snake; others are noble and courageous and high-bred, as the lion; others are thorough-bred and wild and treacherous, as the wolf: for, by the way, an animal is highbred if it come from a noble stock, and an animal is thorough-bred if it does not deflect from its racial characteristics.

Further, some are crafty and mischievous, as the fox; some are spirited and affectionate and fawning, as the dog; others are easy-tempered and easily domesticated, as the elephant; others are cautious and watchful, as the goose; others are jealous and self-conceited, as the peacock. But of all animals man alone is capable of deliberation.

Many animals have memory, and are capable of instruction; but no other creature except man can recall the past at will.

With regard to the several genera of animals, particulars as to their habits of life and modes of existence will be discussed more fully by and by.

Part 2

Common to all animals are the organs whereby they take food and the organs where into they take it; and these are either identical with one another, or are diverse in the ways above specified: to wit, either identical in form, or varying in respect of excess or defect, or resembling one another analogically, or differing in position.

Furthermore, the great majority of animals have other organs besides these in common, whereby they discharge the residuum of their food: I say, the great majority, for this statement does not apply to all. And, by the way, the organ whereby food is taken in is called the mouth, and the organ whereinto it is taken, the belly; the remainder of the alimentary system has a great variety of names.

Now the residuum of food is twofold in kind, wet and dry, and such creatures as have organs receptive of wet residuum are invariably found with organs receptive of dry residuum; but such as have organs receptive of dry residuum need not possess organs receptive of wet residuum. In other words, an animal has a bowel or intestine if it have a bladder; but an animal may have a bowel and be without a bladder. And, by the way, I may here remark that the organ receptive of wet residuum is termed 'bladder', and the organ receptive of dry residuum 'intestine or 'bowel'.

Part 3

Of animals otherwise, a great many have, besides the organs above-mentioned, an organ for excretion of the sperm: and of animals capable of generation one secretes into another, and the other into itself. The latter is termed 'female', and the former 'male'; but some animals have neither male nor female. Consequently, the organs connected with this function differ in form, for some animals have a womb and others an organ analogous thereto.

The above-mentioned organs, then, are the most indispensable parts of animals; and with some of them all animals without exception, and with others animals for the most part, must needs be provided.

One sense, and one alone, is common to all animals-the sense of touch. Consequently, there is no special name for the organ in which it has its seat; for in some groups of animals the organ is identical, in others it is only analogous.

Part 4

Every animal is supplied with moisture, and, if the animal be deprived of the same by natural causes or artificial means, death ensues: further, every animal has another part in which the moisture is contained. These parts are blood and vein, and in other animals there is something to correspond; but in these latter the parts are imperfect, being merely fibre and serum or lymph.

Touch has its seat in a part uniform and homogeneous, as in the flesh or something of the kind, and generally, with animals supplied with blood, in the parts charged with blood. In other animals it has its seat in parts analogous to the parts charged with blood; but in all cases it is seated in parts that in their texture are homogeneous.

The active faculties, on the contrary, are seated in the parts that are heterogeneous: as, for instance, the business of preparing the food is seated in the mouth, and the office of locomotion in the feet, the wings, or in organs to correspond.

Again, some animals are supplied with blood, as man, the horse, and all such animals as are, when full-grown, either destitute of feet, or two-footed, or four-footed; other animals are bloodless, such as the bee and the wasp, and, of marine animals, the cuttle-fish, the crawfish, and all such animals as have more than four feet.

Part 5

Again, some animals are viviparous, others oviparous, others vermiparous or 'grub-bearing'. Some are viviparous, such as man, the horse, the seal, and all other animals that are hair-coated, and, of marine animals, the cetaceans, as the dolphin, and the so-called Selachia. (Of these latter animals, some have a tubular air-passage and no gills, as the dolphin and the whale: the dolphin with the air-passage going through its back, the whale with the air-passage in its forehead; others have uncovered gills, as the Selachia, the sharks and rays.)

What we term an egg is a certain completed result of conception out of which the animal that is to be develops, and in such a way that in respect to its primitive germ it comes from part only of the egg, while the rest serves for food as the germ develops. A 'grub' on the other hand is a thing out of which in its entirety the animal in its entirety develops, by differentiation and growth of the embryo.

Of viviparous animals, some hatch eggs in their own interior, as creatures of the shark kind; others engender in their interior a live fetus, as man and the horse. When the result of conception is perfected, with some animals a living creature is brought forth, with others an egg is brought to light, with others a grub. Of the eggs, some have egg-shells and are of two different colours within, such as birds' eggs; others are soft-skinned and of uniform colour, as the eggs of animals of the shark kind. Of the grubs, some are from the first capable of movement, others are motionless. However, with regard to these phenomena we shall speak precisely hereafter when we come to treat of Generation.

Furthermore, some animals have feet and some are destitute thereof. Of such as have feet some animals have two, as is the case with men and birds, and with men and birds only; some have four, as the lizard and the dog; some have more, as the centipede and the bee; but allsoever that have feet have an even number of them.

Of swimming creatures that are destitute of feet, some have winglets or fins, as fishes: and of these some have four fins, two above on the back, two below on the belly, as the gilthead and the bass; some have two only,-to wit, such as are exceedingly long and smooth, as the eel and the conger; some have none at all, as the muraena, but use the sea just as snakes use dry ground-and by the way, snakes swim in water in just the same way. Of the shark-kind some have no fins, such as those that are flat and long-tailed, as the ray and the sting-ray, but these fishes swim actually by the undulatory motion of their flat bodies; the fishing frog, however, has fins, and so likewise have all such fishes as have not their flat surfaces thinned off to a sharp edge.

Of those swimming creatures that appear to have feet, as is the case with the molluscs, these creatures swim by the aid of their feet and their fins as well, and they swim most rapidly backwards in the direction of the trunk, as is the case with the cuttle-fish or sepia and the calamari; and, by the way, neither of these latter can walk as the poulpe or octopus can.

The hard-skinned or crustaceous animals, like the crawfish, swim by the instrumentality of their tail-parts; and they swim most rapidly tail foremost, by the aid of the fins developed upon that member. The newt swims by means of its feet and tail; and its tail resembles that of the sheatfish, to compare little with great.

Of animals that can fly some are furnished with feathered wings, as the eagle and the hawk; some are furnished with membranous wings, as the bee and the cockchafer; others are furnished with leathern wings, as the flying fox and the bat. All flying creatures possessed of blood have feathered wings or leathern wings; the bloodless creatures have membranous wings, as insects. The creatures that have feathered wings or leathern wings have either two feet or no feet at all: for there are said to be certain flying serpents in Ethiopia that are destitute of feet.

Creatures that have feathered wings are classed as a genus under the name of 'bird'; the other two genera, the leathern-winged and membrane-winged, are as yet without a generic title.

Of creatures that can fly and are bloodless some are coleopterous or sheath-winged, for they have their wings in a sheath or shard, like the cockchafer and the dung-beetle; others are sheathless, and of these latter some are dipterous and some tetrapterous: tetrapterous, such as are comparatively large or have their stings in the tail, dipterous, such as are comparatively small or have their stings in front. The coleoptera are, without exception, devoid of stings; the diptera have the sting in front, as the fly, the horsefly, the gadfly, and the gnat.

Bloodless animals as a general rule are inferior in point of size to blooded animals; though, by the way, there are found in the sea some few bloodless creatures of abnormal size, as in the case of certain molluscs. And of these bloodless genera, those are the largest that dwell in milder climates, and those that inhabit the sea are larger than those living on dry land or in fresh water.

All creatures that are capable of motion move with four or more points of motion; the blooded animals with four only: as, for instance, man with two hands and two feet, birds with two wings and two feet, quadrupeds and fishes severally with four feet and four fins. Creatures that have two winglets or fins, or that have none at all like serpents, move all the same with not less than four points of motion; for there are four bends in their bodies as they move, or two bends together with their fins. Bloodless and many footed animals, whether furnished with wings or feet, move with more than four points of motion; as, for instance, the dayfly moves with four feet and four wings: and, I may observe in passing, this creature is exceptional not only in regard to the duration of its existence, whence it receives its name, but also because though a quadruped it has wings also.

All animals move alike, four-footed and many-footed; in other words, they all move cross-corner-wise. And animals in general have two feet in advance; the crab alone has four.

Part 6

Very extensive genera of animals, into which other subdivisions fall, are the following: one, of birds; one, of fishes; and another, of cetaceans. Now all these creatures are blooded.

There is another genus of the hard-shell kind, which is called oyster; another of the soft-shell kind, not as yet designated by a single term, such as the spiny crawfish and the various kinds of crabs and lobsters; and another of molluscs, as the two kinds of calamari and the cuttle-fish; that of insects is different. All these latter creatures are bloodless, and such of them as have feet have a goodly number of them; and of the insects some have wings as well as feet.

Of the other animals the genera are not extensive. For in them one species does not comprehend many species; but in one case, as man, the species is simple, admitting of no differentiation, while other cases admit of differentiation, but the forms lack particular designations.

So, for instance, creatures that are quadupedal and unprovided with wings are blooded without exception, but some of them are viviparous, and some oviparous. Such as are viviparous are hair-coated, and such as are oviparous are covered with a kind of tessellated hard substance; and the tessellated bits of this substance are, as it were, similar in regard to position to a scale.

An animal that is blooded and capable of movement on dry land, but is naturally unprovided with feet, belongs to the serpent genus; and animals of this genus are coated with the tessellated horny substance. Serpents in general are oviparous; the adder, an exceptional case, is viviparous: for not all viviparous animals are hair-coated, and some fishes also are viviparous.

All animals, however, that are hair-coated are viviparous. For, by the way, one must regard as a kind of hair such prickly hairs as hedgehogs and porcupines carry; for these spines perform the office of hair, and not of feet as is the case with similar parts of sea-urchins.

In the genus that combines all viviparous quadrupeds are many species, but under no common appellation. They are only named as it were one by one, as we say man, lion, stag, horse, dog, and so on; though, by the way, there is a sort of genus that embraces all creatures that have bushy manes and bushy tails, such as the horse, the ass, the mule, the jennet, and the animals that are called Hemioni in Syria,-from their externally resembling mules, though they are not strictly of the same species. And that they are not so is proved by the fact that they mate with and breed from one another. For all these reasons, we must take animals species by species, and discuss their peculiarities severally'

These preceding statements, then, have been put forward thus in a general way, as a kind of foretaste of the number of subjects and of the properties that we have to consider in order that we may first get a clear notion of distinctive character and common properties. By and by we shall discuss these matters with greater minuteness.

After this we shall pass on to the discussion of causes. For to do this when the investigation of the details is complete is the proper and natural method, and that whereby the subjects and the premises of our argument will afterwards be rendered plain.

In the first place we must look to the constituent parts of animals. For it is in a way relative to these parts, first and foremost, that animals in their entirety differ from one another: either in the fact that some have this or that, while they have not that or this; or by peculiarities of position or of arrangement; or by the differences that have been previously mentioned, depending upon diversity of form, or excess or defect in this or that particular, on analogy, or on contrasts of the accidental qualities.

To begin with, we must take into consideration the parts of Man. For, just as each nation is wont to reckon by that monetary standard with which it is most familiar, so must we do in other matters. And, of course, man is the animal with which we are all of us the most familiar.

Now the parts are obvious enough to physical perception. However, with the view of observing due order and sequence and of combining rational notions with physical perception, we shall proceed to enumerate the parts: firstly, the organic, and afterwards the simple or non-composite.

Part 7

The chief parts into which the body as a whole is subdivided, are the head, the neck, the trunk (extending from the neck to the privy parts), which is called the thorax, two arms and two legs.

Of the parts of which the head is composed the hair-covered portion is called the 'skull'. The front portion of it is termed 'bregma' or 'sinciput', developed after birth-for it is the last of all the bones in the body to acquire solidity,-the hinder part is termed the 'occiput', and the part intervening between the sinciput and the occiput is the 'crown'. The brain lies underneath the sinciput; the occiput is hollow. The skull consists entirely of thin bone, rounded in shape, and contained within a wrapper of fleshless skin.

The skull has sutures: one, of circular form, in the case of women; in the case of men, as a general rule, three meeting at a point. Instances have been known of a man's skull devoid of suture altogether. In the skull the middle line, where the hair parts, is called the crown or vertex. In some cases the parting is double; that is to say, some men are double crowned, not in regard to the bony skull, but in consequence of the double fall or set of the hair.

Part 8

The part that lies under the skull is called the 'face': but in the case of man only, for the term is not applied to a fish or to an ox. In the face the part below the sinciput and between the eyes is termed the forehead. When men have large foreheads, they are slow to move; when they have small ones, they are fickle; when they have broad ones, they are apt to be distraught; when they have foreheads rounded or bulging out, they are quick-tempered.

Part 9

Underneath the forehead are two eyebrows. Straight eyebrows are a sign of softness of disposition; such as curve in towards the nose, of harshness; such as curve out towards the temples, of humour and dissimulation; such as are drawn in towards one another, of jealousy.

Under the eyebrows come the eyes. These are naturally two in number. Each of them has an upper and a lower eyelid, and the hairs on the edges of these are termed 'eyelashes'. The central part of the eye includes the moist part whereby vision is effected, termed the 'pupil', and the part surrounding it called the 'black'; the part outside this is the 'white'. A part common to the upper and lower eyelid is a pair of nicks or corners, one in the direction of the nose, and the other in the direction of the temples. When these are long they are a sign of bad disposition; if the side toward the nostril be fleshy and comb-like, they are a sign of dishonesty.

All animals, as a general rule, are provided with eyes, excepting the ostracoderms and other imperfect creatures; at all events, all viviparous animals have eyes, with the exception of the mole. And yet one might assert that, though the mole has not eyes in the full sense, yet it has eyes in a kind of a way. For in point of absolute fact it cannot see, and has no eyes visible externally; but when the outer skin is removed, it is found to have the place where eyes are usually situated, and the black parts of the eyes rightly situated, and all the place that is usually devoted on the outside to eyes: showing that the parts are stunted in development, and the skin allowed to grow over.

Part 10

Of the eye the white is pretty much the same in all creatures; but what is called the black differs in various animals. Some have the rim black, some distinctly blue, some greyish-blue, some greenish; and this last colour is the sign of an excellent disposition, and is particularly well adapted for sharpness of vision. Man is the only, or nearly the only, creature, that has eyes of diverse colours. Animals, as a rule, have eyes of one colour only. Some horses have blue eyes.

Of eyes, some are large, some small, some medium-sized; of these, the medium-sized are the best. Moreover, eyes sometimes protrude, sometimes recede, sometimes are neither protruding nor receding. Of these, the receding eye is in all animals the most acute; but the last kind are the sign of the best disposition. Again, eyes are sometimes inclined to wink under observation, sometimes to remain open and staring, and sometimes are disposed neither to wink nor stare. The last kind are the sign of the best nature, and of the others, the latter kind indicates impudence, and the former indecision.

Part 11

Furthermore, there is a portion of the head, whereby an animal hears, a part incapable of breathing, the 'ear'. I say 'incapable of breathing', for Alcmaeon is mistaken when he says that goats inspire through their ears. Of the ear one part is unnamed, the other part is called the 'lobe'; and it is entirely composed of gristle and flesh. The ear is constructed internally like the trumpet-shell, and the innermost bone is like the ear itself, and into it at the end the sound makes its way, as into the bottom of a jar. This receptacle does not communicate by any passage with the brain, but does so with the palate, and a vein extends from the brain towards it. The eyes also are connected with the brain, and each of them lies at the end of a little vein. Of animals possessed of ears man is the only one that cannot move this organ. Of creatures possessed of hearing, some have ears, whilst others have none, but merely have the passages for ears visible, as, for example, feathered animals or animals coated with horny tessellates.

Viviparous animals, with the exception of the seal, the dolphin, and those others which after a similar fashion to these are cetaceans, are all provided with ears; for, by the way, the shark-kind are also viviparous. Now, the seal has the passages visible whereby it hears; but the dolphin can hear, but has no ears, nor yet any passages visible. But man alone is unable to move his ears, and all other animals can move them. And the ears lie, with man, in the same horizontal plane with the eyes, and not in a plane above them as is the case with some quadrupeds. Of ears, some are fine, some are coarse, and some are of medium texture; the last kind are best for hearing, but they serve in no way to indicate character. Some ears are large, some small, some medium-sized; again, some stand out far, some lie in close and tight, and some take up a medium position; of these such as are of medium size and of medium position are indications of the best disposition, while the large and outstanding ones indicate a tendency to irrelevant talk or chattering. The part intercepted between the eye, the ear, and the crown is termed the 'temple'. Again, there is a part of the countenance that serves as a passage for the breath, the 'nose'. For a man inhales and exhales by this organ, and sneezing is effected by its means: which last is an outward rush of collected breath, and is the only mode of breath used as an omen and regarded as supernatural. Both inhalation and exhalation go right on from the nose towards the chest; and with the nostrils alone and separately it is impossible to inhale or exhale, owing to the fact that the inspiration and respiration take place from the chest along the windpipe, and not by any portion connected with the head; and indeed it is possible for a creature to live without using this process of nasal respiration.

Again, smelling takes place by means of the nose,-smelling, or the sensible discrimination of odour. And the nostril admits of easy motion, and is not, like the ear, intrinsically immovable. A part of it, composed of gristle, constitutes, a septum or partition, and part is an open passage; for the nostril consists of two separate channels. The nostril (or nose) of the elephant is long and strong, and the animal uses it like a hand; for by means of this organ it draws objects towards it, and takes hold of them, and introduces its food into its mouth, whether liquid or dry food, and it is the only living creature that does so.

Furthermore, there are two jaws; the front part of them constitutes the chin, and the hinder part the cheek. All animals move the lower jaw, with the exception of the river crocodile; this creature moves the upper jaw only.

Next after the nose come two lips, composed of flesh, and facile of motion. The mouth lies inside the jaws and lips. Parts of the mouth are the roof or palate and the pharynx.

The part that is sensible of taste is the tongue. The sensation has its seat at the tip of the tongue; if the object to be tasted be placed on the flat surface of the organ, the taste is less sensibly experienced. The tongue is sensitive in all other ways wherein flesh in general is so: that is, it can appreciate hardness, or warmth and cold, in any part of it, just as it can appreciate taste. The tongue is sometimes broad, sometimes narrow, and sometimes of medium width; the last kind is the best and the clearest in its discrimination of taste. Moreover, the tongue is sometimes loosely hung, and sometimes fastened: as in the case of those who mumble and who lisp.

The tongue consists of flesh, soft and spongy, and the so-called 'epiglottis' is a part of this organ.

That part of the mouth that splits into two bits is called the 'tonsils'; that part that splits into many bits, the 'gums'. Both the tonsils and the gums are composed of flesh. In the gums are teeth, composed of bone.

Inside the mouth is another part, shaped like a bunch of grapes, a pillar streaked with veins. If this pillar gets relaxed and inflamed it is called 'uvula' or 'bunch of grapes', and it then has a tendency to bring about suffocation.

Part 12

The neck is the part between the face and the trunk. Of this the front part is the larynx land the back part the ur The front part, composed of gristle, through which respiration and speech is effected, is termed the 'windpipe'; the part that is fleshy is the esophagus, inside just in front of the chine. The part to the back of the neck is the epomis, or 'shoulder-point'.

These then are the parts to be met with before you come to the thorax.

To the trunk there is a front part and a back part. Next after the neck in the front part is the chest, with a pair of breasts. To each of the breasts is attached a teat or nipple, through which in the case of females the milk percolates; and the breast is of a spongy texture. Milk, by the way, is found at times in the male; but with the male the flesh of the breast is tough, with the female it is soft and porous.

Part 13

Next after the thorax and in front comes the 'belly', and its root the 'navel'. Underneath this root the bilateral part is the 'flank': the undivided part below the navel, the 'abdomen', the extremity of which is the region of the 'pubes'; above the navel the 'hypochondrium'; the cavity common to the hypochondrium and the flank is the gut-cavity.

Serving as a brace girdle to the hinder parts is the pelvis, and hence it gets its name (osphus), for it is symmetrical (isophues) in appearance; of the fundament the part for resting on is termed the 'rump', and the part whereon the thigh pivots is termed the 'socket' (or acetabulum).

The 'womb' is a part peculiar to the female; and the 'penis' is peculiar to the male. This latter organ is external and situated at the extremity of the trunk; it is composed of two separate parts: of which the extreme part is fleshy, does not alter in size, and is called the glans; and round about it is a skin devoid of any specific title, which integument if it be cut asunder never grows together again, any more than does the jaw or the eyelid. And the connexion between the latter and the glans is called the frenum. The remaining part of the penis is composed of gristle; it is easily susceptible of enlargement; and it protrudes and recedes in the reverse directions to what is observable in the identical organ in cats. Underneath the penis are two 'testicles', and the integument of these is a skin that is termed the 'scrotum'.

Testicles are not identical with flesh, and are not altogether diverse from it. But by and by we shall treat in an exhaustive way regarding all such parts.

Part 14

The privy part of the female is in character opposite to that of the male. In other words, the part under the pubes is hollow or receding, and not, like the male organ, protruding. Further, there is a 'urethra' outside the womb; which organ serves as a passage for the sperm of the male, and as an outlet for liquid excretion to both sexes).

The part common to the neck and chest is the 'throat'; the 'armpit' is common to side, arm, and shoulder; and the 'groin' is common to thigh and abdomen. The part inside the thigh and buttocks is the 'perineum', and the part outside the thigh and buttocks is the 'hypoglutis'.

The front parts of the trunk have now been enumerated. The part behind the chest is termed the 'back'.

Part 15

Parts of the back are a pair of 'shoulderblades', the 'back-bone', and, underneath on a level with the belly in the trunk, the 'loins'. Common to the upper and lower part of the trunk are the 'ribs', eight on either side, for as to the so-called seven-ribbed Ligyans we have not received any trustworthy evidence.

Man, then, has an upper and a lower part, a front and a back part, a right and a left side. Now the right and the left side are pretty well alike in their parts and identical throughout, except that the left side is the weaker of the two; but the back parts do not resemble the front ones, neither do the lower ones the upper: only that these upper and lower parts may be said to resemble one another thus far, that, if the face be plump or meager, the abdomen is plump or meager to correspond; and that the legs correspond to the arms, and where the upper arm is short the thigh is usually short also, and where the feet are small the hands are small correspondingly.

Of the limbs, one set, forming a pair, is 'arms'. To the arm belong the 'shoulder', 'upper-arm', 'elbow', 'fore-arm', and 'hand'. To the hand belong the 'palm', and the five 'fingers'. The part of the finger that bends is termed 'knuckle', the part that is inflexible is termed the 'phalanx'. The big finger or thumb is single-jointed, the other fingers are double jointed. The bending both of the arm and of the finger takes place from without inwards in all cases; and the arm bends at the elbow. The inner part of the hand is termed the palm', and is fleshy and divided by joints or lines: in the case of long-lived people by one or two extending right across, in the case of the short-lived by two, not so extending. The joint between hand and arm is termed the 'wrist'. The outside or back of the hand is sinewy, and has no specific designation.

There is another duplicate limb, the 'leg'. Of this limb the double-knobbed part is termed the 'thigh-bone', the sliding part of the 'kneecap', the double-boned part the 'leg'; the front part of this latter is termed the 'shin', and the part behind it the 'calf', wherein the flesh is sinewy and venous, in some cases drawn upwards towards the hollow behind the knee, as in the case of people with large hips, and in other cases drawn downwards. The lower extremity of the shin is the 'ankle', duplicate in either leg. The part of the limb that contains a multiplicity of bones is the 'foot'. The hinder part of the foot is the 'heel'; at the front of it the divided part consists of 'toes', five in number; the fleshy part underneath is the 'ball'; the upper part or back of the foot is sinewy and has no particular appellation; of the toe, one portion is the 'nail' and another the 'joint', and the nail is in all cases at the extremity; and toes are without exception single jointed. Men that have the inside or sole of the foot clumsy and not arched, that is, that walk resting on the entire under-surface of their feet, are prone to roguery. The joint common to thigh and shin is the 'knee'.

These, then, are the parts common to the male and the female sex. The relative position of the parts as to up and down, or to front and back, or to right and left, all this as regards externals might safely be left to mere ordinary perception. But for all that, we must treat of them for the same reason as the one previously brought forward; that is to say, we must refer to them in order that a due and regular sequence may be observed in our exposition, and in order that by the enumeration of these obvious facts due attention may be subsequently given to those parts in men and other animals that are diverse in any way from one another.

In man, above all other animals, the terms 'upper' and 'lower' are used in harmony with their natural positions; for in him, upper and lower have the same meaning as when they are applied to the universe as a whole. In like manner the terms, 'in front', 'behind', 'right' and 'left', are used in accordance with their natural sense. But in regard to other animals, in some cases these distinctions do not exist, and in others they do so, but in a vague way. For instance, the head with all animals is up and above in respect to their bodies; but man alone, as has been said, has, in maturity, this part uppermost in respect to the material universe.

Next after the head comes the neck, and then the chest and the back: the one in front and the other behind. Next after these come the belly, the loins, the sexual parts, and the haunches; then the thigh and shin; and, lastly, the feet.

The legs bend frontwards, in the direction of actual progression, and frontwards also lies that part of the foot which is the most effective of motion, and the flexure of that part; but the heel lies at the back, and the anklebones lie laterally, earwise. The arms are situated to right and left, and bend inwards: so that the convexities formed by bent arms and legs are practically face to face with one another in the case of man.

As for the senses and for the organs of sensation, the eyes, the nostrils, and the tongue, all alike are situated frontwards; the sense of hearing, and the organ of hearing, the ear, is situated sideways, on the same horizontal plane with the eyes. The eyes in man are, in proportion to his size, nearer to one another than in any other animal.

Of the senses man has the sense of touch more refined than any animal, and so also, but in less degree, the sense of taste; in the development of the other senses he is surpassed by a great number of animals.

Part 16

The parts, then, that are externally visible are arranged in the way above stated, and as a rule have their special designations, and from use and wont are known familiarly to all; but this is not the case with the inner parts. For the fact is that the inner parts of man are to a very great extent unknown, and the consequence is that we must have recourse to an examination of the inner parts of other animals whose nature in any way resembles that of man.

In the first place then, the brain lies in the front part of the head. And this holds alike with all animals possessed of a brain; and all blooded animals are possessed thereof, and, by the way, molluscs as well. But, taking size for size of animal, the largest brain, and the moistest, is that of man. Two membranes enclose it: the stronger one near the bone of the skull; the inner one, round the brain itself, is finer. The brain in all cases is bilateral. Behind this, right at the back, comes what is termed the 'cerebellum', differing in form from the brain as we may both feel and see.

The back of the head is with all animals empty and hollow, whatever be its size in the different animals. For some creatures have big heads while the face below is small in proportion, as is the case with round-faced animals; some have little heads and long jaws, as is the case, without exception, among animals of the mane-and-tail species.

The brain in all animals is bloodless, devoid of veins, and naturally cold to the touch; in the great majority of animals it has a small hollow in its centre. The brain-caul around it is reticulated with veins; and this brain-caul is that skin-like membrane which closely surrounds the brain. Above the brain is the thinnest and weakest bone of the head, which is termed or 'sinciput'.

From the eye there go three ducts to the brain: the largest and the medium-sized to the cerebellum, the least to the brain itself; and the least is the one situated nearest to the nostril. The two largest ones, then, run side by side and do not meet; the medium-sized ones meet-and this is particularly visible in fishes,-for they lie nearer than the large ones to the brain; the smallest pair are the most widely separate from one another, and do not meet.

Inside the neck is what is termed the esophagus (whose other name is derived esophagus from its length and narrowness), and the windpipe. The windpipe is situated in front of the esophagus in all animals that have a windpipe, and all animals have one that are furnished with lungs. The windpipe is made up of gristle, is sparingly supplied with blood, and is streaked all round with numerous minute veins; it is situated, in its upper part, near the mouth, below the aperture formed by the nostrils into the mouth-an aperture through which, when men, in drinking, inhale any of the liquid, this liquid finds its way out through the nostrils. In betwixt the two openings comes the so-called epiglottis, an organ capable of being drawn over and covering the orifice of the windpipe communicating with the mouth; the end of the tongue is attached to the epiglottis. In the other direction the windpipe extends to the interval between the lungs, and hereupon bifurcates into each of the two divisions of the lung; for the lung in all animals possessed of the organ has a tendency to be double. In viviparous animals, however, the duplication is not so plainly discernible as in other species, and the duplication is least discernible in man. And in man the organ is not split into many parts, as is the case with some vivipara, neither is it smooth, but its surface is uneven.

In the case of the ovipara, such as birds and oviparous quadrupeds, the two parts of the organ are separated to a distance from one another, so that the creatures appear to be furnished with a pair of lungs; and from the windpipe, itself single, there branch off two separate parts extending to each of the two divisions of the lung. It is attached also to the great vein and to what is designated the 'aorta'. When the windpipe is charged with air, the air passes on to the hollow parts of the lung. These parts have divisions, composed of gristle, which meet at an acute angle; from the divisions run passages through the entire lung, giving off smaller and smaller ramifications. The heart also is attached to the windpipe, by connexions of fat, gristle, and sinew; and at the point of juncture there is a hollow. When the windpipe is charged with air, the entrance of the air into the heart, though imperceptible in some animals, is perceptible enough in the larger ones. Such are the properties of the windpipe, and it takes in and throws out air only, and takes in nothing else either dry or liquid, or else it causes you pain until you shall have coughed up whatever may have gone down.

The esophagus communicates at the top with the mouth, close to the windpipe, and is attached to the backbone and the windpipe by membranous ligaments, and at last finds its way through the midriff into the belly. It is composed of flesh-like substance, and is elastic both lengthways and breadthways.

The stomach of man resembles that of a dog; for it is not much bigger than the bowel, but is somewhat like a bowel of more than usual width; then comes the bowel, single, convoluted, moderately wide. The lower part of the gut is like that of a pig; for it is broad, and the part from it to the buttocks is thick and short. The caul, or great omentum, is attached to the middle of the stomach, and consists of a fatty membrane, as is the case with all other animals whose stomachs are single and which have teeth in both jaws.

The mesentery is over the bowels; this also is membranous and broad, and turns to fat. It is attached to the great vein and the aorta, and there run through it a number of veins closely packed together, extending towards the region of the bowels, beginning above and ending below.

So much for the properties of the esophagus, the windpipe, and the stomach.

Part 17

The heart has three cavities, and is situated above the lung at the division of the windpipe, and is provided with a fatty and thick membrane where it fastens on to the great vein and the aorta. It lies with its tapering portion upon the aorta, and this portion is similarly situated in relation to the chest in all animals that have a chest. In all animals alike, in those that have a chest and in those that have none, the apex of the heart points forwards, although this fact might possibly escape notice by a change of position under dissection. The rounded end of the heart is at the top. The apex is to a great extent fleshy and close in texture, and in the cavities of the heart are sinews. As a rule the heart is situated in the middle of the chest in animals that have a chest, and in man it is situated a little to the left-hand side, leaning a little way from the division of the breasts towards the left breast in the upper part of the chest.

The heart is not large, and in its general shape it is not elongated; in fact, it is somewhat round in form: only, be it remembered, it is sharp-pointed at the bottom. It has three cavities, as has been said: the right-hand one the largest of the three, the left-hand one the least, and the middle one intermediate in size. All these cavities, even the two small ones, are connected by passages with the lung, and this fact is rendered quite plain in one of the cavities. And below, at the point of attachment, in the largest cavity there is a connexion with the great vein (near which the mesentery lies); and in the middle one there is a connexion with the aorta.

Canals lead from the heart into the lung, and branch off just as the windpipe does, running all over the lung parallel with the passages from the windpipe. The canals from the heart are uppermost; and there is no common passage, but the passages through their having a common wall receive the breath and pass it on to the heart; and one of the passages conveys it to the right cavity, and the other to the left.

With regard to the great vein and the aorta we shall, by and by, treat of them together in a discussion devoted to them and to them alone. In all animals that are furnished with a lung, and that are both internally and externally viviparous, the lung is of all organs the most richly supplied with blood; for the lung is throughout spongy in texture, and along by every single pore in it go branches from the great vein. Those who imagine it to be empty are altogether mistaken; and they are led into their error by their observation of lungs removed from animals under dissection, out of which organs the blood had all escaped immediately after death.

Of the other internal organs the heart alone contains blood. And the lung has blood not in itself but in its veins, but the heart has blood in itself; for in each of its three cavities it has blood, but the thinnest blood is what it has in its central cavity.

Under the lung comes the thoracic diaphragm or midriff, attached to the ribs, the hypochondria and the backbone, with a thin membrane in the middle of it. It has veins running through it; and the diaphragm in the case of man is thicker in proportion to the size of his frame than in other animals.

Under the diaphragm on the right-hand side lies the 'liver', and on the left-hand side the 'spleen', alike in all animals that are provided with these organs in an ordinary and not preternatural way; for, be it observed, in some quadrupeds these organs have been found in a transposed position. These organs are connected with the stomach by the caul.

To outward view the spleen of man is narrow and long, resembling the self-same organ in the pig. The liver in the great majority of animals is not provided with a 'gall-bladder'; but the latter is present in some. The liver of a man is round-shaped, and resembles the same organ in the ox. And, by the way, the absence above referred to of a gall-bladder is at times met with in the practice of augury. For instance, in a certain district of the Chalcidic settlement in Euboea the sheep are devoid of gall-bladders; and in Naxos nearly all the quadrupeds have one so large that foreigners when they offer sacrifice with such victims are bewildered with fright, under the impression that the phenomenon is not due to natural causes, but bodes some mischief to the individual offerers of the sacrifice.

Again, the liver is attached to the great vein, but it has no communication with the aorta; for the vein that goes off from the great vein goes right through the liver, at a point where are the so-called 'portals' of the liver. The spleen also is connected only with the great vein, for a vein extends to the spleen off from it.

After these organs come the 'kidneys', and these are placed close to the backbone, and resemble in character the same organ in kine. In all animals that are provided with this organ, the right kidney is situated higher up than the other. It has also less fatty substance than the left-hand one and is less moist. And this phenomenon also is observable in all the other animals alike.

Furthermore, passages or ducts lead into the kidneys both from the great vein and from the aorta, only not into the cavity. For, by the way, there is a cavity in the middle of the kidney, bigger in some creatures and less in others; but there is none in the case of the seal. This latter animal has kidneys resembling in shape the identical organ in kine, but in its case the organs are more solid than in any other known creature. The ducts that lead into the kidneys lose themselves in the substance of the kidneys themselves; and the proof that they extend no farther rests on the fact that they contain no blood, nor is any clot found therein. The kidneys, however, have, as has been said, a small cavity. From this cavity in the kidney there lead two considerable ducts or ureters into the bladder; and others spring from the aorta, strong and continuous. And to the middle of each of the two kidneys is attached a hollow sinewy vein, stretching right along the spine through the narrows; by and by these veins are lost in either loin, and again become visible extending to the flank. And these off-branchings of the veins terminate in the bladder. For the bladder lies at the extremity, and is held in position by the ducts stretching from the kidneys, along the stalk that extends to the urethra; and pretty well all round it is fastened by fine sinewy membranes, that resemble to some extent the thoracic diaphragm. The bladder in man is, proportionately to his size, tolerably large.

To the stalk of the bladder the private part is attached, the external orifices coalescing; but a little lower down, one of the openings communicates with the testicles and the other with the bladder. The penis is gristly and sinewy in its texture. With it are connected the testicles in male animals, and the properties of these organs we shall discuss in our general account of the said organ.

All these organs are similar in the female; for there is no difference in regard to the internal organs, except in respect to the womb, and with reference to the appearance of this organ I must refer the reader to diagrams in my 'Anatomy'. The womb, however, is situated over the bowel, and the bladder lies over the womb. But we must treat by and by in our pages of the womb of all female animals viewed generally. For the wombs of all female animals are not identical, neither do their local dispositions coincide.

These are the organs, internal and external, of man, and such is their nature and such their local disposition.

Book II

Part 1

WITH regard to animals in general, some parts or organs are common to all, as has been said, and some are common only to particular genera; the parts, moreover, are identical with or different from one another on the lines already repeatedly laid down. For as a general rule all animals that are generically distinct have the majority of their parts or organs different in form or species; and some of them they have only analogically similar and diverse in kind or genus, while they have others that are alike in kind but specifically diverse; and many parts or organs exist in some animals, but not in others.

For instance, viviparous quadrupeds have all a head and a neck, and all the parts or organs of the head, but they differ each from other in the shapes of the parts. The lion has its neck composed of one single bone instead of vertebrae; but, when dissected, the animal is found in all internal characters to resemble the dog.

The quadrupedal vivipara instead of arms have forelegs. This is true of all quadrupeds, but such of them as have toes have, practically speaking, organs analogous to hands; at all events, they use these fore-limbs for many purposes as hands. And they have the limbs on the left-hand side less distinct from those on the right than man.

The fore-limbs then serve more or less the purpose of hands in quadrupeds, with the exception of the elephant. This latter animal has its toes somewhat indistinctly defined, and its front legs are much bigger than its hinder ones; it is five-toed, and has short ankles to its hind feet. But it has a nose such in properties and such in size as to allow of its using the same for a hand. For it eats and drinks by lifting up its food with the aid of this organ into its mouth, and with the same organ it lifts up articles to the driver on its back; with this organ it can pluck up trees by the roots, and when walking through water it spouts the water up by means of it; and this organ is capable of being crooked or coiled at the tip, but not of flexing like a joint, for it is composed of gristle.

Of all animals man alone can learn to make equal use of both hands.

All animals have a part analogous to the chest in man, but not similar to his; for the chest in man is broad, but that of all other animals is narrow. Moreover, no other animal but man has breasts in front; the elephant, certainly, has two breasts, not however in the chest, but near it.

Moreover, also, animals have the flexions of their fore and hind limbs in directions opposite to one another, and in directions the reverse of those observed in the arms and legs of man; with the exception of the elephant. In other words, with the viviparous quadrupeds the front legs bend forwards and the hind ones backwards, and the concavities of the two pairs of limbs thus face one another.

The elephant does not sleep standing, as some were wont to assert, but it bends its legs and settles down; only that in consequence of its weight it cannot bend its leg on both sides simultaneously, but falls into a recumbent position on one side or the other, and in this position it goes to sleep. And it bends its hind legs just as a man bends his legs.

In the case of the ovipara, as the crocodile and the lizard and the like, both pairs of legs, fore and hind, bend forwards, with a slight swerve on one side. The flexion is similar in the case of the multipeds; only that the legs in between the extreme ends always move in a manner intermediate between that of those in front and those behind, and accordingly bend sideways rather than backwards or forwards. But man bends his arms and his legs towards the same point, and therefore in opposite ways: that is to say, he bends his arms backwards, with just a slight inclination inwards, and his legs frontwards. No animal bends both its fore-limbs and hind-limbs backwards; but in the case of all animals the flexion of the shoulders is in the opposite direction to that of the elbows or the joints of the forelegs, and the flexure in the hips to that of the knees of the hind-legs: so that since man differs from other animals in flexion, those animals that possess such parts as these move them contrariwise to man.

Birds have the flexions of their limbs like those of the quadrupeds; for, although bipeds, they bend their legs backwards, and instead of arms or front legs have wings which bend frontwards.

The seal is a kind of imperfect or crippled quadruped; for just behind the shoulder-blade its front feet are placed, resembling hands, like the front paws of the bear; for they are furnished with five toes, and each of the toes has three flexions and a nail of inconsiderable size. The hind feet are also furnished with five toes; in their flexions and nails they resemble the front feet, and in shape they resemble a fish's tail.

The movements of animals, quadruped and multiped, are crosswise, or in diagonals, and their equilibrium in standing posture is maintained crosswise; and it is always the limb on the right-hand side that is the first to move. The lion, however, and the two species of camels, both the Bactrian and the Arabian, progress by an amble; and the action so called is when the animal never overpasses the right with the left, but always follows close upon it.

Whatever parts men have in front, these parts quadrupeds have below, in or on the belly; and whatever parts men have behind, these parts quadrupeds have above on their backs. Most quadrupeds have a tail; for even the seal has a tiny one resembling that of the stag. Regarding the tails of the pithecoids we must give their distinctive properties by and by animal

All viviparous quadrupeds are hair-coated, whereas man has only a few short hairs excepting on the head, but, so far as the head is concerned, he is hairier than any other animal. Further, of hair-coated animals, the back is hairier than the belly, which latter is either comparatively void of hair or smooth and void of hair altogether. With man the reverse is the case.

Man also has upper and lower eyelashes, and hair under the armpits and on the pubes. No other animal has hair in either of these localities, or has an under eyelash; though in the case of some animals a few straggling hairs grow under the eyelid.

Of hair-coated quadrupeds some are hairy all over the body, as the pig, the bear, and the dog; others are especially hairy on the neck and all round about it, as is the case with animals that have a shaggy mane, such as the lion; others again are especially hairy on the upper surface of the neck from the head as far as the withers, namely, such as have a crested mane, as in the case with the horse, the mule, and, among the undomesticated horned animals, the bison.

The so-called hippelaphus also has a mane on its withers, and the animal called pardion, in either case a thin mane extending from the head to the withers; the hippelaphus has, exceptionally, a beard by the larynx. Both these animals have horns and are cloven-footed; the female, however, of the hippelaphus has no horns. This latter animal resembles the stag in size; it is found in the territory of the Arachotae, where the wild cattle also are found. Wild cattle differ from their domesticated congeners just as the wild boar differs from the domesticated one. That is to say they are black, strong looking, with a hook-nosed muzzle, and with horns lying more over the back. The horns of the hippelaphus resemble those of the gazelle.

The elephant, by the way, is the least hairy of all quadrupeds. With animals, as a general rule, the tail corresponds with the body as regards thickness or thinness of hair-coating; that is, with animals that have long tails, for some creatures have tails of altogether insignificant size.

Camels have an exceptional organ wherein they differ from all other animals, and that is the so-called 'hump' on their back. The Bactrian camel differs from the Arabian; for the former has two humps and the latter only one, though it has, by the way, a kind of a hump below like the one above, on which, when it kneels, the weight of the whole body rests. The camel has four teats like the cow, a tail like that of an ass, and the privy parts of the male are directed backwards. It has one knee in each leg, and the flexures of the limb are not manifold, as some say, although they appear to be so from the constricted shape of the region of the belly. It has a huckle-bone like that of kine, but meager and small in proportion to its bulk. It is cloven-footed, and has not got teeth in both jaws; and it is cloven footed in the following way: at the back there is a slight cleft extending as far up as the second joint of the toes; and in front there are small hooves on the tip of the first joint of the toes; and a sort of web passes across the cleft, as in geese. The foot is fleshy underneath, like that of the bear; so that, when the animal goes to war, they protect its feet, when they get sore, with sandals.

The legs of all quadrupeds are bony, sinewy, and fleshless; and in point of fact such is the case with all animals that are furnished with feet, with the exception of man. They are also unfurnished with buttocks; and this last point is plain in an especial degree in birds. It is the reverse with man; for there is scarcely any part of the body in which man is so fleshy as in the buttock, the thigh, and the calf; for the part of the leg called gastroenemia or is fleshy.

Of blooded and viviparous quadrupeds some have the foot cloven into many parts, as is the case with the hands and feet of man (for some animals, by the way, are many-toed, as the lion, the dog, and the pard); others have feet cloven in twain, and instead of nails have hooves, as the sheep, the goat, the deer, and the hippopotamus; others are uncloven of foot, such for instance as the solid-hooved animals, the horse and the mule. Swine are either cloven-footed or uncloven-footed; for there are in Illyria and in Paeonia and elsewhere solid-hooved swine. The cloven-footed animals have two clefts behind; in the solid-hooved this part is continuous and undivided.

Furthermore, of animals some are horned, and some are not so. The great majority of the horned animals are cloven-footed, as the ox, the stag, the goat; and a solid-hooved animal with a pair of horns has never yet been met with. But a few animals are known to be singled-horned and single-hooved, as the Indian ass; and one, to wit the oryx, is single horned and cloven-hooved.

Of all solid-hooved animals the Indian ass alone has an astragalus or huckle-bone; for the pig, as was said above, is either solid-hooved or cloven-footed, and consequently has no well-formed huckle-bone. Of the cloven footed many are provided with a huckle-bone. Of the many-fingered or many-toed, no single one has been observed to have a huckle-bone, none of the others any more than man. The lynx, however, has something like a hemiastragal, and the lion something resembling the sculptor's 'labyrinth'. All the animals that have a huckle-bone have it in the hinder legs. They have also the bone placed straight up in the joint; the upper part, outside; the lower part, inside; the sides called Coa turned towards one another, the sides called Chia outside, and the keraiae or 'horns' on the top. This, then, is the position of the hucklebone in the case of all animals provided with the part.

Some animals are, at one and the same time, furnished with a mane and furnished also with a pair of horns bent in towards one another, as is the bison (or aurochs), which is found in Paeonia and Maedica. But all animals that are horned are quadrupedal, except in cases where a creature is said metaphorically, or by a figure of speech, to have horns; just as the Egyptians describe the serpents found in the neighbourhood of Thebes, while in point of fact the creatures have merely protuberances on the head sufficiently large to suggest such an epithet.

Of horned animals the deer alone has a horn, or antler, hard and solid throughout. The horns of other animals are hollow for a certain distance, and solid towards the extremity. The hollow part is derived from the skin, but the core round which this is wrapped-the hard part-is derived from the bones; as is the case with the horns of oxen. The deer is the only animal that sheds its horns, and it does so annually, after reaching the age of two years, and again renews them. All other animals retain their horns permanently, unless the horns be damaged by accident.

Again, with regard to the breasts and the generative organs, animals differ widely from one another and from man. For instance, the breasts of some animals are situated in front, either in the chest or near to it, and there are in such cases two breasts and two teats, as is the case with man and the elephant, as previously stated. For the elephant has two breasts in the region of the axillae; and the female elephant has two breasts insignificant in size and in no way proportionate to the bulk of the entire frame, in fact, so insignificant as to be invisible in a sideways view; the males also have breasts, like the females, exceedingly small. The she-bear has four breasts. Some animals have two breasts, but situated near the thighs, and teats, likewise two in number, as the sheep; others have four teats, as the cow. Some have breasts neither in the chest nor at the thighs, but in the belly, as the dog and pig; and they have a considerable number of breasts or dugs, but not all of equal size. Thus the leopard has four dugs in the belly, the lioness two, and others more. The she-camel, also, has two dugs and four teats, like the cow. Of solid-hooved animals the males have no dugs, excepting in the case of males that take after the mother, which phenomenon is observable in horses.

Of male animals the genitals of some are external, as is the case with man, the horse, and most other creatures; some are internal, as with the dolphin. With those that have the organ externally placed, the organ in some cases is situated in front, as in the cases already mentioned, and of these some have the organ detached, both penis and testicles, as man; others have penis and testicles closely attached to the belly, some more closely, some less; for this organ is not detached in the wild boar nor in the horse.

The penis of the elephant resembles that of the horse; compared with the size of the animal it is disproportionately small; the testicles are not visible, but are concealed inside in the vicinity of the kidneys; and for this reason the male speedily gives over in the act of intercourse. The genitals of the female are situated where the udder is in sheep; when she is in heat, she draws the organ back and exposes it externally, to facilitate the act of intercourse for the male; and the organ opens out to a considerable extent.

With most animals the genitals have the position above assigned; but some animals discharge their urine backwards, as the lynx, the lion, the camel, and the hare. Male animals differ from one another, as has been said, in this particular, but all female animals are retromingent: even the female elephant like other animals, though she has the privy part below the thighs.

In the male organ itself there is a great diversity. For in some cases the organ is composed of flesh and gristle, as in man; in such cases, the fleshy part does not become inflated, but the gristly part is subject to enlargement. In other cases, the organ is composed of fibrous tissue, as with the camel and the deer; in other cases it is bony, as with the fox, the wolf, the marten, and the weasel; for this organ in the weasel has a bone.

When man has arrived at maturity, his upper part is smaller than the lower one, but with all other blooded animals the reverse holds good. By the 'upper' part we mean all extending from the head down to the parts used for excretion of residuum, and by the 'lower' part else. With animals that have feet the hind legs are to be rated as the lower part in our comparison of magnitudes, and with animals devoid of feet, the tail, and the like.

When animals arrive at maturity, their properties are as above stated; but they differ greatly from one another in their growth towards maturity. For instance, man, when young, has his upper part larger than the lower, but in course of growth he comes to reverse this condition; and it is owing to this circumstance that-an exceptional instance, by the way-he does not progress in early life as he does at maturity, but in infancy creeps on all fours; but some animals, in growth, retain the relative proportion of the parts, as the dog. Some animals at first have the upper part smaller and the lower part larger, and in course of growth the upper part gets to be the larger, as is the case with the bushy-tailed animals such as the horse; for in their case there is never, subsequently to birth, any increase in the part extending from the hoof to the haunch.

Again, in respect to the teeth, animals differ greatly both from one another and from man. All animals that are quadrupedal, blooded, and viviparous, are furnished with teeth; but, to begin with, some are double-toothed (or fully furnished with teeth in both jaws), and some are not. For instance, horned quadrupeds are not double-toothed; for they have not got the front teeth in the upper jaw; and some hornless animals, also, are not double toothed, as the camel. Some animals have tusks, like the boar, and some have not. Further, some animals are saw-toothed, such as the lion, the pard, and the dog; and some have teeth that do not interlock but have flat opposing crowns, as the horse and the ox; and by 'saw-toothed' we mean such animals as interlock the sharp-pointed teeth in one jaw between the sharp-pointed ones in the other. No animal is there that possesses both tusks and horns, nor yet do either of these structures exist in any animal possessed of 'saw-teeth'. The front teeth are usually sharp, and the back ones blunt. The seal is saw-toothed throughout, inasmuch as he is a sort of link with the class of fishes; for fishes are almost all saw-toothed.

No animal of these genera is provided with double rows of teeth. There is, however, an animal of the sort, if we are to believe Ctesias. He assures us that the Indian wild beast called the 'martichoras' has a triple row of teeth in both upper and lower jaw; that it is as big as a lion and equally hairy, and that its feet resemble those of the lion; that it resembles man in its face and ears; that its eyes are blue, and its colour vermilion; that its tail is like that of the land-scorpion; that it has a sting in the tail, and has the faculty of shooting off arrow-wise the spines that are attached to the tail; that the sound of its voice is a something between the sound of a pan-pipe and that of a trumpet; that it can run as swiftly as deer, and that it is savage and a man-eater.

Man sheds his teeth, and so do other animals, as the horse, the mule, and the ass. And man sheds his front teeth; but there is no instance of an animal that sheds its molars. The pig sheds none of its teeth at all.

Part 2

With regard to dogs some doubts are entertained, as some contend that they shed no teeth whatever, and others that they shed the canines, but those alone; the fact being, that they do shed their teeth like man, but that the circumstance escapes observation, owing to the fact that they never shed them until equivalent teeth have grown within the gums to take the place of the shed ones. We shall be justified in supposing that the case is similar with wild beasts in general; for they are said to shed their canines only. Dogs can be distinguished from one another, the young from the old, by their teeth; for the teeth in young dogs are white and sharp-pointed; in old dogs, black and blunt.

Part 3

In this particular, the horse differs entirely from animals in general: for, generally speaking, as animals grow older their teeth get blacker, but the horse's teeth grow whiter with age.

The so-called 'canines' come in between the sharp teeth and the broad or blunt ones, partaking of the form of both kinds; for they are broad at the base and sharp at the tip.

Males have more teeth than females in the case of men, sheep, goats, and swine; in the case of other animals observations have not yet been made: but the more teeth they have the more long-lived are they, as a rule, while those are short-lived in proportion that have teeth fewer in number and thinly set.

Part 4

The last teeth to come in man are molars called 'wisdom-teeth', which come at the age of twenty years, in the case of both sexes. Cases have been known in women upwards. of eighty years old where at the very close of life the wisdom-teeth have come up, causing great pain in their coming; and cases have been known of the like phenomenon in men too. This happens, when it does happen, in the case of people where the wisdom-teeth have not come up in early years.

Part 5

The elephant has four teeth on either side, by which it munches its food, grinding it like so much barley-meal, and, quite apart from these, it has its great teeth, or tusks, two in number. In the male these tusks are comparatively large and curved upwards; in the female, they are comparatively small and point in the opposite direction; that is, they look downwards towards the ground. The elephant is furnished with teeth at birth, but the tusks are not then visible.

Part 6

The tongue of the elephant is exceedingly small, and situated far back in the mouth, so that it is difficult to get a sight of it.

Part 7

Furthermore, animals differ from one another in the relative size of their mouths. In some animals the mouth opens wide, as is the case with the dog, the lion, and with all the saw-toothed animals; other animals have small mouths, as man; and others have mouths of medium capacity, as the pig and his congeners.

(The Egyptian hippopotamus has a mane like a horse, is cloven-footed like an ox, and is snub-nosed. It has a huckle-bone like cloven-footed animals, and tusks just visible; it has the tail of a pig, the neigh of a horse, and the dimensions of an ass. The hide is so thick that spears are made out of it. In its internal organs it resembles the horse and the ass.)

Part 8

Some animals share the properties of man and the quadrupeds, as the ape, the monkey, and the baboon. The monkey is a tailed ape. The baboon resembles the ape in form, only that it is bigger and stronger, more like a dog in face, and is more savage in its habits, and its teeth are more dog-like and more powerful.

Apes are hairy on the back in keeping with their quadrupedal nature, and hairy on the belly in keeping with their human form-for, as was said above, this characteristic is reversed in man and the quadruped-only that the hair is coarse, so that the ape is thickly coated both on the belly and on the back. Its face resembles that of man in many respects; in other words, it has similar nostrils and ears, and teeth like those of man, both front teeth and molars. Further, whereas quadrupeds in general are not furnished with lashes on one of the two eyelids, this creature has them on both, only very thinly set, especially the under ones; in fact they are very insignificant indeed. And we must bear in mind that all other quadrupeds have no under eyelash at all.

The ape has also in its chest two teats upon poorly developed breasts. It has also arms like man, only covered with hair, and it bends these legs like man, with the convexities of both limbs facing one another. In addition, it has hands and fingers and nails like man, only that all these parts are somewhat more beast-like in appearance. Its feet are exceptional in kind. That is, they are like large hands, and the toes are like fingers, with the middle one the longest of all, and the under part of the foot is like a hand except for its length, and stretches out towards the extremities like the palm of the hand; and this palm at the after end is unusually hard, and in a clumsy obscure kind of way resembles a heel. The creature uses its feet either as hands or feet, and doubles them up as one doubles a fist. Its upper-arm and thigh are short in proportion to the forearm and the shin. It has no projecting navel, but only a hardness in the ordinary locality of the navel. Its upper part is much larger than its lower part, as is the case with quadrupeds; in fact, the proportion of the former to the latter is about as five to three. Owing to this circumstance and to the fact that its feet resemble hands and are composed in a manner of hand and of foot: of foot in the heel extremity, of the hand in all else-for even the toes have what is called a 'palm':-for these reasons the animal is oftener to be found on all fours than upright. It has neither hips, inasmuch as it is a quadruped, nor yet a tail, inasmuch as it is a biped, except nor yet a tail by the way that it has a tail as small as small can be, just a sort of indication of a tail. The genitals of the female resemble those of the female in the human species; those of the male are more like those of a dog than are those of a man.

Part 9

The monkey, as has been observed, is furnished with a tail. In all such creatures the internal organs are found under dissection to correspond to those of man.

So much then for the properties of the organs of such animals as bring forth their young into the world alive.

Part 10

Oviparous and blooded quadrupeds-and, by the way, no terrestrial blooded animal is oviparous unless it is quadrupedal or is devoid of feet altogether-are furnished with a head, a neck, a back, upper and under parts, the front legs and hind legs, and the part analogous to the chest, all as in the case of viviparous quadrupeds, and with a tail, usually large, in exceptional cases small. And all these creatures are many-toed, and the several toes are cloven apart. Furthermore, they all have the ordinary organs of sensation, including a tongue, with the exception of the Egyptian crocodile.

This latter animal, by the way, resembles certain fishes. For, as a general rule, fishes have a prickly tongue, not free in its movements; though there are some fishes that present a smooth undifferentiated surface where the tongue should be, until you open their mouths wide and make a close inspection.

Again, oviparous blooded quadrupeds are unprovided with ears, but possess only the passage for hearing; neither have they breasts, nor a copulatory organ, nor external testicles, but internal ones only; neither are they hair coated, but are in all cases covered with scaly plates. Moreover, they are without exception saw-toothed.

River crocodiles have pigs' eyes, large teeth and tusks, and strong nails, and an impenetrable skin composed of scaly plates. They see but poorly under water, but above the surface of it with remarkable acuteness. As a rule, they pass the day-time on land and the nighttime in the water; for the temperature of the water is at night-time more genial than that of the open air.

Part 11

The chameleon resembles the lizard in the general configuration of its body, but the ribs stretch downwards and meet together under the belly as is the case with fishes, and the spine sticks up as with the fish. Its face resembles that of the baboon. Its tail is exceedingly long, terminates in a sharp point, and is for the most part coiled up, like a strap of leather. It stands higher off the ground than the lizard, but the flexure of the legs is the same in both creatures. Each of its feet is divided into two parts, which bear the same relation to one another that the thumb and the rest of the hand bear to one another in man. Each of these parts is for a short distance divided after a fashion into toes; on the front feet the inside part is divided into three and the outside into two, on the hind feet the inside part into two and the outside into three; it has claws also on these parts resembling those of birds of prey. Its body is rough all over, like that of the crocodile. Its eyes are situated in a hollow recess, and are very large and round, and are enveloped in a skin resembling that which covers the entire body; and in the middle a slight aperture is left for vision, through which the animal sees, for it never covers up this aperture with the cutaneous envelope. It keeps twisting its eyes round and shifting its line of vision in every direction, and thus contrives to get a sight of any object that it wants to see. The change in its colour takes place when it is inflated with air; it is then black, not unlike the crocodile, or green like the lizard but black-spotted like the pard. This change of colour takes place over the whole body alike, for the eyes and the tail come alike under its influence. In its movements it is very sluggish, like the tortoise. It assumes a greenish hue in dying, and retains this hue after death. It resembles the lizard in the position of the esophagus and the windpipe. It has no flesh anywhere except a few scraps of flesh on the head and on the jaws and near to the root of the tail. It has blood only round about the heart, the eyes, the region above the heart, and in all the veins extending from these parts; and in all these there is but little blood after all. The brain is situated a little above the eyes, but connected with them. When the outer skin is drawn aside from off the eye, a something is found surrounding the eye, that gleams through like a thin ring of copper. Membranes extend well nigh over its entire frame, numerous and strong, and surpassing in respect of number and relative strength those found in any other animal. After being cut open along its entire length it continues to breathe for a considerable time; a very slight motion goes on in the region of the heart, and, while contraction is especially manifested in the neighbourhood of the ribs, a similar motion is more or less discernible over the whole body. It has no spleen visible. It hibernates, like the lizard.

Part 12

Birds also in some parts resemble the above mentioned animals; that is to say, they have in all cases a head, a neck, a back, a belly, and what is analogous to the chest. The bird is remarkable among animals as having two feet, like man; only, by the way, it bends them backwards as quadrupeds bend their hind legs, as was noticed previously. It has neither hands nor front feet, but wings-an exceptional structure as compared with other animals. Its haunch-bone is long, like a thigh, and is attached to the body as far as the middle of the belly; so like to a thigh is it that when viewed separately it looks like a real one, while the real thigh is a separate structure betwixt it and the shin. Of all birds those that have crooked talons have the biggest thighs and the strongest breasts. All birds are furnished with many claws, and all have the toes separated more or less asunder; that is to say, in the greater part the toes are clearly distinct from one another, for even the swimming birds, although they are web-footed, have still their claws fully articulated and distinctly differentiated from one another. Birds that fly high in air are in all cases four-toed: that is, the greater part have three toes in front and one behind in place of a heel; some few have two in front and two behind, as the wryneck.

This latter bird is somewhat bigger than the chaffinch, and is mottled in appearance. It is peculiar in the arrangement of its toes, and resembles the snake in the structure of its tongue; for the creature can protrude its tongue to the extent of four finger-breadths, and then draw it back again. Moreover, it can twist its head backwards while keeping all the rest of its body still, like the serpent. It has big claws, somewhat resembling those of the woodpecker. Its note is a shrill chirp.

Birds are furnished with a mouth, but with an exceptional one, for they have neither lips nor teeth, but a beak. Neither have they ears nor a nose, but only passages for the sensations connected with these organs: that for the nostrils in the beak, and that for hearing in the head. Like all other animals they all have two eyes, and these are devoid of lashes. The heavy-bodied (or gallinaceous) birds close the eye by means of the lower lid, and all birds blink by means of a skin extending over the eye from the inner corner; the owl and its congeners also close the eye by means of the upper lid. The same phenomenon is observable in the animals that are protected by horny scutes, as in the lizard and its congeners; for they all without exception close the eye with the lower lid, but they do not blink like birds. Further, birds have neither scutes nor hair, but feathers; and the feathers are invariably furnished with quills. They have no tail, but a rump with tail-feathers, short in such as are long-legged and web-footed, large in others. These latter kinds of birds fly with their feet tucked up close to the belly; but the small rumped or short-tailed birds fly with their legs stretched out at full length. All are furnished with a tongue, but the organ is variable, being long in some birds and broad in others. Certain species of birds above all other animals, and next after man, possess the faculty of uttering articulate sounds; and this faculty is chiefly developed in broad-tongued birds. No oviparous creature has an epiglottis over the windpipe, but these animals so manage the opening and shutting of the windpipe as not to allow any solid substance to get down into the lung.

Some species of birds are furnished additionally with spurs, but no bird with crooked talons is found so provided. The birds with talons are among those that fly well, but those that have spurs are among the heavy-bodied.

Again, some birds have a crest. As a general rule the crest sticks up, and is composed of feathers only; but the crest of the barn-door cock is exceptional in kind, for, whereas it is not just exactly flesh, at the same time it is not easy to say what else it is.

Part 13

Of water animals the genus of fishes constitutes a single group apart from the rest, and including many diverse forms.

In the first place, the fish has a head, a back, a belly, in the neighbourhood of which last are placed the stomach and viscera; and behind it has a tail of continuous, undivided shape, but not, by the way, in all cases alike. No fish has a neck, or any limb, or testicles at all, within or without, or breasts. But, by the way this absence of breasts may predicated of all non-viviparous animals; and in point of fact viviparous animals are not in all cases provided with the organ, excepting such as are directly viviparous without being first oviparous. Thus the dolphin is directly viviparous, and accordingly we find it furnished with two breasts, not situated high up, but in the neighbourhood of the genitals. And this creature is not provided, like quadrupeds, with visible teats, but has two vents, one on each flank, from which the milk flows; and its young have to follow after it to get suckled, and this phenomenon has been actually witnessed.

Fishes, then, as has been observed, have no breasts and no passage for the genitals visible externally. But they have an exceptional organ in the gills, whereby, after taking the water in the mouth, they discharge it again; and in the fins, of which the greater part have four, and the lanky ones two, as, for instance, the eel, and these two situated near to the gills. In like manner the grey mullet-as, for instance, the mullet found in the lake at Siphae-have only two fins; and the same is the case with the fish called Ribbon-fish. Some of the lanky fishes have no fins at all, such as the muraena, nor gills articulated like those of other fish.

And of those fish that are provided with gills, some have coverings for this organ, whereas all the selachians have the organ unprotected by a cover. And those fishes that have coverings or opercula for the gills have in all cases their gills placed sideways; whereas, among selachians, the broad ones have the gills down below on the belly, as the torpedo and the ray, while the lanky ones have the organ placed sideways, as is the case in all the dog-fish.

The fishing-frog has gills placed sideways, and covered not with a spiny operculum, as in all but the selachian fishes, but with one of skin.

Moreover, with fishes furnished with gills, the gills in some cases are simple in others duplicate; and the last gill in the direction of the body is always simple. And, again, some fishes have few gills, and others have a great number; but all alike have the same number on both sides. Those that have the least number have one gill on either side, and this one duplicate, like the boar-fish; others have two on either side, one simple and the other duplicate, like the conger and the scarus; others have four on either side, simple, as the elops, the synagris, the muraena, and the eel; others have four, all, with the exception of the hindmost one, in double rows, as the wrasse, the perch, the sheat-fish, and the carp. The dog-fish have all their gills double, five on a side; and the sword-fish has eight double gills. So much for the number of gills as found in fishes.

Again, fishes differ from other animals in more ways than as regards the gills. For they are not covered with hairs as are viviparous land animals, nor, as is the case with certain oviparous quadrupeds, with tessellated scutes, nor, like birds, with feathers; but for the most part they are covered with scales. Some few are rough-skinned, while the smooth-skinned are very few indeed. Of the Selachia some are rough-skinned and some smooth-skinned; and among the smooth-skinned fishes are included the conger, the eel, and the tunny.

All fishes are saw-toothed excepting the scarus; and the teeth in all cases are sharp and set in many rows, and in some cases are placed on the tongue. The tongue is hard and spiny, and so firmly attached that fishes in many instances seem to be devoid of the organ altogether. The mouth in some cases is wide-stretched, as it is with some viviparous quadrupeds....

With regard to organs of sense, all save eyes, fishes possess none of them, neither the organs nor their passages, neither ears nor nostrils; but all fishes are furnished with eyes, and the eyes devoid of lids, though the eyes are not hard; with regard to the organs connected with the other senses, hearing and smell, they are devoid alike of the organs themselves and of passages indicative of them.

Fishes without exception are supplied with blood. Some of them are oviparous, and some viviparous; scaly fish are invariably oviparous, but cartilaginous fishes are all viviparous, with the single exception of the fishing-frog.

Part 14

Of blooded animals there now remains the serpent genus. This genus is common to both elements, for, while most species comprehended therein are land animals, a small minority, to wit the aquatic species, pass their lives in fresh water. There are also sea-serpents, in shape to a great extent resembling their congeners of the land, with this exception that the head in their case is somewhat like the head of the conger; and there are several kinds of sea-serpent, and the different kinds differ in colour; these animals are not found in very deep water. Serpents, like fish, are devoid of feet.

There are also sea-scolopendras, resembling in shape their land congeners, but somewhat less in regard to magnitude. These creatures are found in the neighbourhood of rocks; as compared with their land congeners they are redder in colour, are furnished with feet in greater numbers and with legs of more delicate structure. And the same remark applies to them as to the sea-serpents, that they are not found in very deep water.

Of fishes whose habitat is in the vicinity of rocks there is a tiny one, which some call the Echeneis, or 'ship-holder', and which is by some people used as a charm to bring luck in affairs of law and love. The creature is unfit for eating. Some people assert that it has feet, but this is not the case: it appears, however, to be furnished with feet from the fact that its fins resemble those organs.

So much, then, for the external parts of blooded animals, as regards their numbers, their properties, and their relative diversities.

Part 15

As for the properties of the internal organs, these we must first discuss in the case of the animals that are supplied with blood. For the principal genera differ from the rest of animals, in that the former are supplied with blood and the latter are not; and the former include man, viviparous and oviparous quadrupeds, birds, fishes, cetaceans, and all the others that come under no general designation by reason of their not forming genera, but groups of which simply the specific name is predicable, as when we say 'the serpent,' the 'crocodile'.

All viviparous quadrupeds, then, are furnished with an esophagus and a windpipe, situated as in man; the same statement is applicable to oviparous quadrupeds and to birds, only that the latter present diversities in the shapes of these organs. As a general rule, all animals that take up air and breathe it in and out are furnished with a lung, a windpipe, and an esophagus, with the windpipe and esophagus not admitting of diversity in situation but admitting of diversity in properties, and with the lung admitting of diversity in both these respects. Further, all blooded animals have a heart and a diaphragm or midriff; but in small animals the existence of the latter organ is not so obvious owing to its delicacy and minute size.

In regard to the heart there is an exceptional phenomenon observable in oxen. In other words, there is one species of ox where, though not in all cases, a bone is found inside the heart. And, by the way, the horse's heart also has a bone inside it.

The genera referred to above are not in all cases furnished with a lung: for instance, the fish is devoid of the organ, as is also every animal furnished with gills. All blooded animals are furnished with a liver. As a general rule blooded animals are furnished with a spleen; but with the great majority of non-viviparous but oviparous animals the spleen is so small as all but to escape observation; and this is the case with almost all birds, as with the pigeon, the kite, the falcon, the owl: in point of fact, the aegocephalus is devoid of the organ altogether. With oviparous quadrupeds the case is much the same as with the viviparous; that is to say, they also have the spleen exceedingly minute, as the tortoise, the freshwater tortoise, the toad, the lizard, the crocodile, and the frog.

Some animals have a gall-bladder close to the liver, and others have not. Of viviparous quadrupeds the deer is without the organ, as also the roe, the horse, the mule, the ass, the seal, and some kinds of pigs. Of deer those that are called Achainae appear to have gall in their tail, but what is so called does resemble gall in colour, though it is not so completely fluid, and the organ internally resembles a spleen.

However, without any exception, stags are found to have maggots living inside the head, and the habitat of these creatures is in the hollow underneath the root of the tongue and in the neighbourhood of the vertebra to which the head is attached. These creatures are as large as the largest grubs; they grow all together in a cluster, and they are usually about twenty in number.

Deer then, as has been observed, are without a gall-bladder; their gut, however, is so bitter that even hounds refuse to eat it unless the animal is exceptionally fat. With the elephant also the liver is unfurnished with a gall-bladder, but when the animal is cut in the region where the organ is found in animals furnished with it, there oozes out a fluid resembling gall, in greater or less quantities. Of animals that take in sea-water and are furnished with a lung, the dolphin is unprovided with a gall-bladder. Birds and fishes all have the organ, as also oviparous quadrupeds, all to a greater or a lesser extent. But of fishes some have the organ close to the liver, as the dogfishes, the sheat-fish, the rhine or angel-fish, the smooth skate, the torpedo, and, of the lanky fishes, the eel, the pipe-fish, and the hammer-headed shark. The callionymus, also, has the gall-bladder close to the liver, and in no other fish does the organ attain so great a relative size. Other fishes have the organ close to the gut, attached to the liver by certain extremely fine ducts. The bonito has the gall-bladder stretched alongside the gut and equaling it in length, and often a double fold of it. others have the organ in the region of the gut; in some cases far off, in others near; as the fishing-frog, the elops, the synagris, the muraena, and the sword-fish. Often animals of the same species show this diversity of position; as, for instance, some congers are found with the organ attached close to the liver, and others with it detached from and below it. The case is much the same with birds: that is, some have the gall-bladder close to the stomach, and others close to the gut, as the pigeon, the raven, the quail, the swallow, and the sparrow; some have it near at once to the liver and to the stomach as the aegocephalus; others have it near at once to the liver and the gut, as the falcon and the kite.

Part 16

Again, all viviparous quadrupeds are furnished with kidneys and a bladder. Of the ovipara that are not quadrupedal there is no instance known of an animal, whether fish or bird, provided with these organs. Of the ovipara that are quadrupedal, the turtle alone is provided with these organs of a magnitude to correspond with the other organs of the animal. In the turtle the kidney resembles the same organ in the ox; that is to say, it looks one single organ composed of a number of small ones. (The bison also resembles the ox in all its internal parts).

Part 17

With all animals that are furnished with these parts, the parts are similarly situated, and with the exception of man, the heart is in the middle; in man, however, as has been observed, the heart is placed a little to the left-hand side. In all animals the pointed end of the heart turns frontwards; only in fish it would at first sight seem otherwise, for the pointed end is turned not towards the breast, but towards the head and the mouth. And (in fish) the apex is attached to a tube just where the right and left gills meet together. There are other ducts extending from the heart to each of the gills, greater in the greater fish, lesser in the lesser; but in the large fishes the duct at the pointed end of the heart is a tube, white-coloured and exceedingly thick. Fishes in some few cases have an esophagus, as the conger and the eel; and in these the organ is small.

In fishes that are furnished with an undivided liver, the organ lies entirely on the right side; where the liver is cloven from the root, the larger half of the organ is on the right side: for in some fishes the two parts are detached from one another, without any coalescence at the root, as is the case with the dogfish. And there is also a species of hare in what is named the Fig district, near Lake Bolbe, and elsewhere, which animal might be taken to have two livers owing to the length of the connecting ducts, similar to the structure in the lung of birds.

The spleen in all cases, when normally placed, is on the left-hand side, and the kidneys also lie in the same position in all creatures that possess them. There have been known instances of quadrupeds under dissection, where the spleen was on the right hand and the liver on the left; but all such cases are regarded as supernatural.

In all animals the wind-pipe extends to the lung, and the manner how, we shall discuss hereafter; and the esophagus, in all that have the organ, extends through the midriff into the stomach. For, by the way, as has been observed, most fishes have no esophagus, but the stomach is united directly with the mouth, so that in some cases when big fish are pursuing little ones, the stomach tumbles forward into the mouth.

All the afore-mentioned animals have a stomach, and one similarly situated, that is to say, situated directly under the midriff; and they have a gut connected therewith and closing at the outlet of the residuum and at what is termed the 'rectum'. However, animals present diversities in the structure of their stomachs. In the first place, of the viviparous quadrupeds, such of the horned animals as are not equally furnished with teeth in both jaws are furnished with four such chambers. These animals, by the way, are those that are said to chew the cud. In these animals the esophagus extends from the mouth downwards along the lung, from the midriff to the big stomach (or paunch); and this stomach is rough inside and semi-partitioned. And connected with it near to the entry of the esophagus is what from its appearance is termed the 'reticulum' (or honeycomb bag); for outside it is like the stomach, but inside it resembles a netted cap; and the reticulum is a great deal smaller than the stomach. Connected with this is the 'echinus' (or many-plies), rough inside and laminated, and of about the same size as the reticulum. Next after this comes what is called the 'enystrum' (or abomasum), larger an longer than the echinus, furnished inside with numerous folds or ridges, large and smooth. After all this comes the gut.

Such is the stomach of those quadrupeds that are horned and have an unsymmetrical dentition; and these animals differ one from another in the shape and size of the parts, and in the fact of the esophagus reaching the stomach central-wise in some cases and sideways in others. Animals that are furnished equally with teeth in both jaws have one stomach; as man, the pig, the dog, the bear, the lion, the wolf. (The Thos, by the by, has all its internal organs similar to the wolf's.)

All these, then have a single stomach, and after that the gut; but the stomach in some is comparatively large, as in the pig and bear, and the stomach of the pig has a few smooth folds or ridges; others have a much smaller stomach, not much bigger than the gut, as the lion, the dog, and man. In the other animals the shape of the stomach varies in the direction of one or other of those already mentioned; that is, the stomach in some animals resembles that of the pig; in others that of the dog, alike with the larger animals and the smaller ones. In all these animals diversities occur in regard to the size, the shape, the thickness or the thinness of the stomach, and also in regard to the place where the esophagus opens into it.

There is also a difference in structure in the gut of the two groups of animals above mentioned (those with unsymmetrical and those with symmetrical dentition) in size, in thickness, and in foldings.

The intestines in those animals whose jaws are unequally furnished with teeth are in all cases the larger, for the animals themselves are larger than those in the other category; for very few of them are small, and no single one of the horned animals is very small. And some possess appendages (or caeca) to the gut, but no animal that has not incisors in both jaws has a straight gut.

The elephant has a gut constricted into chambers, so constructed that the animal appears to have four stomachs; in it the food is found, but there is no distinct and separate receptacle. Its viscera resemble those of the pig, only that the liver is four times the size of that of the ox, and the other viscera in like proportion, while the spleen is comparatively small.

Much the same may be predicated of the properties of the stomach and the gut in oviparous quadrupeds, as in the land tortoise, the turtle, the lizard, both crocodiles, and, in fact, in all animals of the like kind; that is to say, their stomach is one and simple, resembling in some cases that of the pig, and in other cases that of the dog.

The serpent genus is similar and in almost all respects furnished similarly to the saurians among land animals, if one could only imagine these saurians to be increased in length and to be devoid of legs. That is to say, the serpent is coated with tessellated scutes, and resembles the saurian in its back and belly; only, by the way, it has no testicles, but, like fishes, has two ducts converging into one, and an ovary long and bifurcate. The rest of its internal organs are identical with those of the saurians, except that, owing to the narrowness and length of the animal, the viscera are correspondingly narrow and elongated, so that they are apt to escape recognition from the similarities in shape. Thus, the windpipe of the creature is exceptionally long, and the esophagus is longer still, and the windpipe commences so close to the mouth that the tongue appears to be underneath it; and the windpipe seems to project over the tongue, owing to the fact that the tongue draws back into a sheath and does not remain in its place as in other animals. The tongue, moreover, is thin and long and black, and can be protruded to a great distance. And both serpents and saurians have this altogether exceptional property in the tongue, that it is forked at the outer extremity, and this property is the more marked in the serpent, for the tips of his tongue are as thin as hairs. The seal, also, by the way, has a split tongue.

The stomach of the serpent is like a more spacious gut, resembling the stomach of the dog; then comes the gut, long, narrow, and single to the end. The heart is situated close to the pharynx, small and kidney-shaped; and for this reason the organ might in some cases appear not to have the pointed end turned towards the breast. Then comes the lung, single, and articulated with a membranous passage, very long, and quite detached from the heart. The liver is long and simple; the spleen is short and round: as is the case in both respects with the saurians. Its gall resembles that of the fish; the water-snakes have it beside the liver, and the other snakes have it usually beside the gut. These creatures are all saw-toothed. Their ribs are as numerous as the days of the month; in other words, they are thirty in number.

Some affirm that the same phenomenon is observable with serpents as with swallow chicks; in other words, they say that if you prick out a serpent's eyes they will grow again. And further, the tails of saurians and of serpents, if they be cut off, will grow again.

With fishes the properties of the gut and stomach are similar; that is, they have a stomach single and simple, but variable in shape according to species. For in some cases the stomach is gut-shaped, as with the scarus, or parrot-fish; which fish, by the way, appears to be the only fish that chews the cud. And the whole length of the gut is simple, and if it have a reduplication or kink it loosens out again into a simple form.

An exceptional property in fishes and in birds for the most part is the being furnished with gut-appendages or caeca. Birds have them low down and few in number. Fishes have them high up about the stomach, and sometimes numerous, as in the goby, the galeos, the perch, the scorpaena, the citharus, the red mullet, and the sparus; the cestreus or grey mullet has several of them on one side of the belly, and on the other side only one. Some fish possess these appendages but only in small numbers, as the hepatus and the glaucus; and, by the way, they are few also in the dorado. These fishes differ also from one another within the same species, for in the dorado one individual has many and another few. Some fishes are entirely without the part, as the majority of the selachians. As for all the rest, some of them have a few and some a great many. And in all cases where the gut-appendages are found in fish, they are found close up to the stomach.

In regard to their internal parts birds differ from other animals and from one another. Some birds, for instance, have a crop in front of the stomach, as the barn-door cock, the cushat, the pigeon, and the partridge; and the crop consists of a large hollow skin, into which the food first enters and where it lies ingested. Just where the crop leaves the esophagus it is somewhat narrow; by and by it broadens out, but where it communicates with the stomach it narrows down again. The stomach (or gizzard) in most birds is fleshy and hard, and inside is a strong skin which comes away from the fleshy part. Other birds have no crop, but instead of it an esophagus wide and roomy, either all the way or in the part leading to the stomach, as with the daw, the raven, and the carrion-crow. The quail also has the esophagus widened out at the lower extremity, and in the aegocephalus and the owl the organ is slightly broader at the bottom than at the top. The duck, the goose, the gull, the catarrhactes, and the great bustard have the esophagus wide and roomy from one end to the other, and the same applies to a great many other birds. In some birds there is a portion of the stomach that resembles a crop, as in the kestrel. In the case of small birds like the swallow and the sparrow neither the esophagus nor the crop is wide, but the stomach is long. Some few have neither a crop nor a dilated esophagus, but the latter is exceedingly long, as in long necked birds, such as the porphyrio, and, by the way, in the case of all these birds the excrement is unusually moist. The quail is exceptional in regard to these organs, as compared with other birds; in other words, it has a crop, and at the same time its esophagus is wide and spacious in front of the stomach, and the crop is at some distance, relatively to its size, from the esophagus at that part.

Further, in most birds, the gut is thin, and simple when loosened out. The gut-appendages or caeca in birds, as has been observed, are few in number, and are not situated high up, as in fishes, but low down towards the extremity of the gut. Birds, then, have caeca-not all, but the greater part of them, such as the barn-door cock, the partridge, the duck, the night-raven, (the localus,) the ascalaphus, the goose, the swan, the great bustard, and the owl. Some of the little birds also have these appendages; but the caeca in their case are exceedingly minute, as in the sparrow.

Book III

Part 1

NOW that we have stated the magnitudes, the properties, and the relative differences of the other internal organs, it remains for us to treat of the organs that contribute to generation. These organs in the female are in all cases internal; in the male they present numerous diversities.

In the blooded animals some males are altogether devoid of testicles, and some have the organ but situated internally; and of those males that have the organ internally situated, some have it close to the loin in the neighbourhood of the kidney and others close to the belly. Other males have the organ situated externally. In the case of these last, the penis is in some cases attached to the belly, whilst in others it is loosely suspended, as is the case also with the testicles; and, in the cases where the penis is attached to the belly, the attachment varies accordingly as the animal is emprosthuretic or opisthuretic.

No fish is furnished with testicles, nor any other creature that has gills, nor any serpent whatever: nor, in short, any animal devoid of feet, save such only as are viviparous within themselves. Birds are furnished with testicles, but these are internally situated, close to the loin. The case is similar with oviparous quadrupeds, such as the lizard, the tortoise and the crocodile; and among the viviparous animals this peculiarity is found in the hedgehog. Others among those creatures that have the organ internally situated have it close to the belly, as is the case with the dolphin amongst animals devoid of feet, and with the elephant among viviparous quadrupeds. In other cases these organs are externally conspicuous.

We have already alluded to the diversities observed in the attachment of these organs to the belly and the adjacent region; in other words, we have stated that in some cases the testicles are tightly fastened back, as in the pig and its allies, and that in others they are freely suspended, as in man.

Fishes, then, are devoid of testicles, as has been stated, and serpents also. They are furnished, however, with two ducts connected with the midriff and running on to either side of the backbone, coalescing into a single duct above the outlet of the residuum, and by 'above' the outlet I mean the region near to the spine. These ducts in the rutting season get filled with the genital fluid, and, if the ducts be squeezed, the sperm oozes out white in colour. As to the differences observed in male fishes of diverse species, the reader should consult my treatise on Anatomy, and the subject will be hereafter more fully discussed when we describe the specific character in each case.

The males of oviparous animals, whether biped or quadruped, are in all cases furnished with testicles close to the loin underneath the midriff. With some animals the organ is whitish, in others somewhat of a sallow hue; in all cases it is entirely enveloped with minute and delicate veins. From each of the two testicles extends a duct, and, as in the case of fishes, the two ducts coalesce into one above the outlet of the residuum. This constitutes the penis, which organ in the case of small ovipara is inconspicuous; but in the case of the larger ovipara, as in the goose and the like, the organ becomes quite visible just after copulation.

The ducts in the case of fishes and in biped and quadruped ovipara are attached to the loin under the stomach and the gut, in betwixt them and the great vein, from which ducts or blood-vessels extend, one to each of the two testicles. And just as with fishes the male sperm is found in the seminal ducts, and the ducts become plainly visible at the rutting season and in some instances become invisible after the season is passed, so also is it with the testicles of birds; before the breeding season the organ is small in some birds and quite invisible in others, but during the season the organ in all cases is greatly enlarged. This phenomenon is remarkably illustrated in the ring-dove and the partridge, so much so that some people are actually of opinion that these birds are devoid of the organ in the winter-time.

Of male animals that have their testicles placed frontwards, some have them inside, close to the belly, as the dolphin; some have them outside, exposed to view, close to the lower extremity of the belly. These animals resemble one another thus far in respect to this organ; but they differ from one another in this fact, that some of them have their testicles situated separately by themselves, while others, which have the organ situated externally, have them enveloped in what is termed the scrotum.

Again, in all viviparous animals furnished with feet the following properties are observed in the testicles themselves. From the aorta there extend vein-like ducts to the head of each of the testicles, and another two from the kidneys; these two from the kidneys are supplied with blood, while the two from the aorta are devoid of it. From the head of the testicle alongside of the testicle itself is a duct, thicker and more sinewy than the other just alluded to-a duct that bends back again at the end of the testicle to its head; and from the head of each of the two testicles the two ducts extend until they coalesce in front at the penis. The duct that bends back again and that which is in contact with the testicle are enveloped in one and the same membrane, so that, until you draw aside the membrane, they present all the appearance of being a single undifferentiated duct. Further, the duct in contact with the testicle has its moist content qualified by blood, but to a comparatively less extent than in the case of the ducts higher up which are connected with the aorta; in the ducts that bend back towards the tube of the penis, the liquid is white-coloured. There also runs a duct from the bladder, opening into the upper part of the canal, around which lies, sheathwise, what is called the 'penis'.

All these descriptive particulars may be regarded by the light of the accompanying diagram; wherein the letter A marks the starting-point of the ducts that extend from the aorta; the letters

KK mark the heads of the testicles and the ducts descending thereunto; the ducts extending from these along the testicles are marked MM; the ducts turning back, in which is the white fluid, are marked BB; the penis D; the bladder E; and the testicles XX.

(By the way, when the testicles are cut off or removed, the ducts draw upwards by contraction. Moreover, when male animals are young, their owner sometimes destroys the organ in them by attrition; sometimes they castrate them at a later period. And I may here add, that a bull has been known to serve a cow immediately after castration, and actually to impregnate her.)

So much then for the properties of testicles in male animals. In female animals furnished with a womb, the womb is not in all cases the same in form or endowed with the same properties, but both in the vivipara and the ovipara great diversities present themselves. In all creatures that have the womb close to the genitals, the womb is two-horned, and one horn lies to the right-hand side and the other to the left; its commencement, however, is single, and so is the orifice, resembling in the case of the most numerous and largest animals a tube composed of much flesh and gristle. Of these parts one is termed the hystera or delphys, whence is derived the word adelphos, and the other part, the tube or orifice, is termed metra. In all biped or quadruped vivipara the womb is in all cases below the midriff, as in man, the dog, the pig, the horse, and the ox; the same is the case also in all horned animals. At the extremity of the so-called ceratia, or horns, the wombs of most animals have a twist or convolution.

In the case of those ovipara that lay eggs externally, the wombs are not in all cases similarly situated. Thus the wombs of birds are close to the midriff, and the wombs of fishes down below, just like the wombs of biped and quadruped vivipara, only that, in the case of the fish, the wombs are delicately formed, membranous, and elongated; so much so that in extremely small fish, each of the two bifurcated parts looks like a single egg, and those fishes whose egg is described as crumbling would appear to have inside them a pair of eggs, whereas in reality each of the two sides consists not of one but of many eggs, and this accounts for their breaking up into so many particles.

The womb of birds has the lower and tubular portion fleshy and firm, and the part close to the midriff membranous and exceedingly thin and fine: so thin and fine that the eggs might seem to be outside the womb altogether. In the larger birds the membrane is more distinctly visible, and, if inflated through the tube, lifts and swells out; in the smaller birds all these parts are more indistinct.

The properties of the womb are similar in oviparous quadrupeds, as the tortoise, the lizard, the frog and the like; for the tube below is single and fleshy, and the cleft portion with the eggs is at the top close to the midriff. With animals devoid of feet that are internally oviparous and viviparous externally, as is the case with the dogfish and the other so-called Selachians (and by this title we designate such creatures destitute of feet and furnished with gills as are viviparous), with these animals the womb is bifurcate, and beginning down below it extends as far as the midriff, as in the case of birds. There is also a narrow part between the two horns running up as far as the midriff, and the eggs are engendered here and above at the origin of the midriff; afterwards they pass into the wider space and turn from eggs into young animals. However, the differences in respect to the wombs of these fishes as compared with others of their own species or with fishes in general, would be more satisfactorily studied in their various forms in specimens under dissection.

The members of the serpent genus also present divergences either when compared with the above-mentioned creatures or with one another. Serpents as a rule are oviparous, the viper being the only viviparous member of the genus. The viper is, previously to external parturition, oviparous internally; and owing to this peculiarity the properties of the womb in the viper are similar to those of the womb in the selachians. The womb of the serpent is long, in keeping with the body, and starting below from a single duct extends continuously on both sides of the spine, so as to give the impression of thus being a separate duct on each side of the spine, until it reaches the midriff, where the eggs are engendered in a row; and these eggs are laid not one by one, but all strung together. (And all animals that are viviparous both internally and externally have the womb situated above the stomach, and all the ovipara underneath, near to the loin. Animals that are viviparous externally and internally oviparous present an intermediate arrangement; for the underneath portion of the womb, in which the eggs are, is placed near to the loin, but the part about the orifice is above the gut.)

Further, there is the following diversity observable in wombs as compared with one another: namely that the females of horned non-ambidental animals are furnished with cotyledons in the womb when they are pregnant, and such is the case, among ambidentals, with the hare, the mouse, and the bat; whereas all other animals that are ambidental, viviparous, and furnished with feet, have the womb quite smooth, and in their case the attachment of the embryo is to the womb itself and not to any cotyledon inside it.

The parts, then, in animals that are not homogeneous with themselves and uniform in their texture, both parts external and parts internal, have the properties above assigned to them.

Part 2

In sanguineous animals the homogeneous or uniform part most universally found is the blood, and its habitat the vein; next in degree of universality, their analogues, lymph and fibre, and, that which chiefly constitutes the frame of animals, flesh and whatsoever in the several parts is analogous to flesh; then bone, and parts that are analogous to bone, as fish-bone and gristle; and then, again, skin, membrane, sinew, hair, nails, and whatever corresponds to these; and, furthermore, fat, suet, and the excretions: and the excretions are dung, phlegm, yellow bile, and black bile.

Now, as the nature of blood and the nature of the veins have all the appearance of being primitive, we must discuss their properties first of all, and all the more as some previous writers have treated them very unsatisfactorily. And the cause of the ignorance thus manifested is the extreme difficulty experienced in the way of observation. For in the dead bodies of animals the nature of the chief veins is undiscoverable, owing to the fact that they collapse at once when the blood leaves them; for the blood pours out of them in a stream, like liquid out of a vessel, since there is no blood separately situated by itself, except a little in the heart, but it is all lodged in the veins. In living animals it is impossible to inspect these parts, for of their very nature they are situated inside the body and out of sight. For this reason anatomists who have carried on their investigations on dead bodies in the dissecting room have failed to discover the chief roots of the veins, while those who have narrowly inspected bodies of living men reduced to extreme attenuation have arrived at conclusions regarding the origin of the veins from the manifestations visible externally. Of these investigators, Syennesis, the physician of Cyprus, writes as follows:-

'The big veins run thus:-from the navel across the loins, along the back, past the lung, in under the breasts; one from right to left, and the other from left to right; that from the left, through the liver to the kidney and the testicle, that from the right, to the spleen and kidney and testicle, and from thence to the penis.' Diogenes of Apollonia writes thus:-

'The veins in man are as follows:-There are two veins pre-eminent in magnitude. These extend through the belly along the backbone, one to right, one to left; either one to the leg on its own side, and upwards to the head, past the collar bones, through the throat. From these, veins extend all over the body, from that on the right hand to the right side and from that on the left hand to the left side; the most important ones, two in number, to the heart in the region of the backbone; other two a little higher up through the chest in underneath the armpit, each to the hand on its side: of these two, one being termed the vein splenitis, and the other the vein hepatitis. Each of the pair splits at its extremity; the one branches in the direction of the thumb and the other in the direction of the palm; and from these run off a number of minute veins branching off to the fingers and to all parts of the hand. Other veins, more minute, extend from the main veins; from that on the right towards the liver, from that on the left towards the spleen and the kidneys. The veins that run to the legs split at the juncture of the legs with the trunk and extend right down the thigh. The largest of these goes down the thigh at the back of it, and can be discerned and traced as a big one; the second one runs inside the thigh, not quite as big as the one just mentioned. After this they pass on along the knee to the shin and the foot (as the upper veins were described as passing towards the hands), and arrive at the sole of the foot, and from thence continue to the toes. Moreover, many delicate veins separate off from the great veins towards the stomach and towards the ribs.

'The veins that run through the throat to the head can be discerned and traced in the neck as large ones; and from each one of the two, where it terminates, there branch off a number of veins to the head; some from the right side towards the left, and some from the left side towards the right; and the two veins terminate near to each of the two ears. There is another pair of veins in the neck running along the big vein on either side, slightly less in size than the pair just spoken of, and with these the greater part of the veins in the head are connected. This other pair runs through the throat inside; and from either one of the two there extend veins in underneath the shoulder blade and towards the hands; and these appear alongside the veins splenitis and hepatitis as another pair of veins smaller in size. When there is a pain near the surface of the body, the physician lances these two latter veins; but when the pain is within and in the region of the stomach he lances the veins splenitis and hepatitis. And from these, other veins depart to run below the breasts.

'There is also another pair running on each side through the spinal marrow to the testicles, thin and delicate. There is, further, a pair running a little underneath the cuticle through the flesh to the kidneys, and these with men terminate at the testicle, and with women at the womb. These veins are termed the spermatic veins. The veins that leave the stomach are comparatively broad just as they leave; but they become gradually thinner, until they change over from right to left and from left to right.

'Blood is thickest when it is imbibed by the fleshy parts; when it is transmitted to the organs above-mentioned, it becomes thin, warm, and frothy.'

Part 3

Such are the accounts given by Syennesis and Diogenes. Polybus writes to the following effect:-

'There are four pairs of veins. The first extends from the back of the head, through the neck on the outside, past the backbone on either side, until it reaches the loins and passes on to the legs, after which it goes on through the shins to the outer side of the ankles and on to the feet. And it is on this account that surgeons, for pains in the back and loin, bleed in the ham and in the outer side of the ankle. Another pair of veins runs from the head, past ears, through the neck; which veins are termed the jugular veins. This pair goes on inside along the backbone, past the muscles of the loins, on to the testicles, and onwards to the thighs, and through the inside of the hams and through the shins down to the inside of the ankles and to the feet; and for this reason, surgeons, for pains in the muscles of the loins and in the testicles, bleed on the hams and the inner side of the ankles. The third pair extends from the temples, through the neck, in underneath the shoulder-blades, into the lung; those from right to left going in underneath the breast and on to the spleen and the kidney; those from left to right running from the lung in underneath the breast and into the liver and the kidney; and both terminate in the fundament. The fourth pair extend from the front part of the head and the eyes in underneath the neck and the collar-bones; from thence they stretch on through the upper part of the upper arms to the elbows and then through the fore-arms on to the wrists and the jointings of the fingers, and also through the lower part of the upper-arms to the armpits, and so on, keeping above the ribs, until one of the pair reaches the spleen and the other reaches the liver; and after this they both pass over the stomach and terminate at the penis.'

The above quotations sum up pretty well the statements of all previous writers. Furthermore, there are some writers on Natural History who have not ventured to lay down the law in such precise terms as regards the veins, but who all alike agree in assigning the head and the brain as the starting-point of the veins. And in this opinion they are mistaken.

The investigation of such a subject, as has been remarked, is one fraught with difficulties; but, if anyone be keenly interested in the matter, his best plan will be to allow his animals to starve to emaciation, then to strangle them on a sudden, and thereupon to prosecute his investigations.

We now proceed to give particulars regarding the properties and functions of the veins. There are two blood-vessels in the thorax by the backbone, and lying to its inner side; and of these two the larger one is situated to the front, and the lesser one is to the rear of it; and the larger is situated rather to the right hand side of the body, and the lesser one to the left; and by some this vein is termed the 'aorta', from the fact that even in dead bodies part of it is observed to be full of air. These blood-vessels have their origins in the heart, for they traverse the other viscera, in whatever direction they happen to run, without in any way losing their distinctive characteristic as blood-vessels, whereas the heart is as it were a part of them (and that too more in respect to the frontward and larger one of the two), owing to the fact that these two veins are above and below, with the heart lying midway.

The heart in all animals has cavities inside it. In the case of the smaller animals even the largest of the chambers is scarcely discernible; the second larger is scarcely discernible in animals of medium size; but in the largest animals all three chambers are distinctly seen. In the heart then (with its pointed end directed frontwards, as has been observed) the largest of the three chambers is on the right-hand side and highest up; the least one is on the left-hand side; and the medium-sized one lies in betwixt the other two; and the largest one of the three chambers is a great deal larger than either of the two others. All three, however, are connected with passages leading in the direction of the lung, but all these communications are indistinctly discernible by reason of their minuteness, except one.

The great blood-vessel, then, is attached to the biggest of the three chambers, the one that lies uppermost and on the right-hand side; it then extends right through the chamber, coming out as blood-vessel again; just as though the cavity of the heart were a part of the vessel, in which the blood broadens its channel as a river that widens out in a lake. The aorta is attached to the middle chamber; only, by the way, it is connected with it by much narrower pipe.

The great blood-vessel then passes through the heart (and runs from the heart into the aorta). The great vessel looks as though made of membrane or skin, while the aorta is narrower than it, and is very sinewy; and as it stretches away to the head and to the lower parts it becomes exceedingly narrow and sinewy.

First of all, then, upwards from the heart there stretches a part of the great blood-vessel towards the lung and the attachment of the aorta, a part consisting of a large undivided vessel. But there split off from it two parts; one towards the lung and the other towards the backbone and the last vertebra of the neck.

The vessel, then, that extends to the lung, as the lung itself is duplicate, divides at first into two; and then extends along by every pipe and every perforation, greater along the greater ones, lesser along the less, so continuously that it is impossible to discern a single part wherein there is not perforation and vein; for the extremities are indistinguishable from their minuteness, and in point of fact the whole lung appears to be filled with blood.

The branches of the blood-vessels lie above the tubes that extend from the windpipe. And that vessel which extends to the vertebra of the neck and the backbone, stretches back again along the backbone; as Homer represents in the lines:-

(Antilochus, as Thoon turned him round), Transpierc'd his back with a dishonest wound; The hollow vein that to the neck extends, Along the chine, the eager javelin rends.

From this vessel there extend small blood-vessels at each rib and each vertebra; and at the vertebra above the kidneys the vessel bifurcates. And in the above way the parts branch off from the great blood-vessel.

But up above all these, from that part which is connected with the heart, the entire vein branches off in two directions. For its branches extend to the sides and to the collarbones, and then pass on, in men through the armpits to the arms, in quadrupeds to the forelegs, in birds to the wings, and in fishes to the upper or pectoral fins. (See diagram.) The trunks of these veins, where they first branch off, are called the 'jugular' veins; and, where they branch off to the neck the great vein run alongside the windpipe; and, occasionally, if these veins are pressed externally, men, though not actually choked, become insensible, shut their eyes, and fall flat on the ground. Extending in the way described and keeping the windpipe in betwixt them, they pass on until they reach the ears at the junction of the lower jaw with the skull. Hence again they branch off into four veins, of which one bends back and descends through the neck and the shoulder, and meets the previous branching off of the vein at the bend of the arm, while the rest of it terminates at the hand and fingers. (See diagram.)

Each vein of the other pair stretches from the region of the ear to the brain, and branches off in a number of fine and delicate veins into the so-called meninx, or membrane, which surrounds the brain. The brain itself in all animals is destitute of blood, and no vein, great or small, holds its course therein. But of the remaining veins that branch off from the last mentioned vein some envelop the head, others close their courses in the organs of sense and at the roots of the teeth in veins exceedingly fine and minute.

Part 4

And in like manner the parts of the lesser one of the two chief blood-vessels, designated the aorta, branch off, accompanying the branches from the big vein; only that, in regard to the aorta, the passages are less in size, and the branches very considerably less than are those of the great vein. So much for the veins as observed in the regions above the heart.

The part of the great vein that lies underneath the heart extends, freely suspended, right through the midriff, and is united both to the aorta and the backbone by slack membranous communications. From it one vein, short and wide, extends through the liver, and from it a number of minute veins branch off into the liver and disappear. From the vein that passes through the liver two branches separate off, of which one terminates in the diaphragm or so-called midriff, and the other runs up again through the armpit into the right arm and unites with the other veins at the inside of the bend of the arm; and it is in consequence of this local connexion that, when the surgeon opens this vein in the forearm, the patient is relieved of certain pains in the liver; and from the left-hand side of it there extends a short but thick vein to the spleen and the little veins branching off it disappear in that organ. Another part branches off from the left-hand side of the great vein, and ascends, by a course similar to the course recently described, into the left arm; only that the ascending vein in the one case is the vein that traverses the liver, while in this case it is distinct from the vein that runs into the spleen. Again, other veins branch off from the big vein; one to the omentum, and another to the pancreas, from which vein run a number of veins through the mesentery. All these veins coalesce in a single large vein, along the entire gut and stomach to the esophagus; about these parts there is a great ramification of branch veins.

As far as the kidneys, each of the two remaining undivided, the aorta and the big vein extend; and here they get more closely attached to the backbone, and branch off, each of the two, into a A shape, and the big vein gets to the rear of the aorta. But the chief attachment of the aorta to the backbone takes place in the region of the heart; and the attachment is effected by means of minute and sinewy vessels. The aorta, just as it draws off from the heart, is a tube of considerable volume, but, as it advances in its course, it gets narrower and more sinewy. And from the aorta there extend veins to the mesentery just like the veins that extend thither from the big vein, only that the branches in the case of the aorta are considerably less in magnitude; they are, indeed, narrow and fibrillar, and they end in delicate hollow fibre-like veinlets.

There is no vessel that runs from the aorta into the liver or the spleen.

From each of the two great blood-vessels there extend branches to each of the two flanks, and both branches fasten on to the bone. Vessels also extend to the kidneys from the big vein and the aorta; only that they do not open into the cavity of the organ, but their ramifications penetrate into its substance. From the aorta run two other ducts to the bladder, firm and continuous; and there are other ducts from the hollow of the kidneys, in no way communicating with the big vein. From the centre of each of the two kidneys springs a hollow sinewy vein, running along the backbone right through the loins; by and by each of the two veins first disappears in its own flank, and soon afterwards reappears stretching in the direction of the flank. The extremities of these attach to the bladder, and also in the male to the penis and in the female to the womb. From the big vein no vein extends to the womb, but the organ is connected with the aorta by veins numerous and closely packed.

Furthermore, from the aorta and the great vein at the points of divarication there branch off other veins. Some of these run to the groins-large hollow veins-and then pass on down through the legs and terminate in the feet and toes. And, again, another set run through the groins and the thighs cross-garter fashion, from right to left and from left to right, and unite in the hams with the other veins.

In the above description we have thrown light upon the course of the veins and their points of departure.

In all sanguineous animals the case stands as here set forth in regard to the points of departure and the courses of the chief veins. But the description does not hold equally good for the entire vein-system in all these animals. For, in point of fact, the organs are not identically situated in them all; and, what is more, some animals are furnished with organs of which other animals are destitute. At the same time, while the description so far holds good, the proof of its accuracy is not equally easy in all cases, but is easiest in the case of animals of considerable magnitude and supplied abundantly with blood. For in little animals and those scantily supplied with blood, either from natural and inherent causes or from a prevalence of fat in the body, thorough accuracy in investigation is not equally attainable; for in the latter of these creatures the passages get clogged, like water-channels choked with slush; and the others have a few minute fibres to serve instead of veins. But in all cases the big vein is plainly discernible, even in creatures of insignificant size.

Part 5

The sinews of animals have the following properties. For these also the point of origin is the heart; for the heart has sinews within itself in the largest of its three chambers, and the aorta is a sinew-like vein; in fact, at its extremity it is actually a sinew, for it is there no longer hollow, and is stretched like the sinews where they terminate at the jointings of the bones. Be it remembered, however, that the sinews do not proceed in unbroken sequence from one point of origin, as do the blood-vessels.

For the veins have the shape of the entire body, like a sketch of a mannequin; in such a way that the whole frame seems to be filled up with little veins in attenuated subjects-for the space occupied by flesh in fat individuals is filled with little veins in thin ones-whereas the sinews are distributed about the joints and the flexures of the bones. Now, if the sinews were derived in unbroken sequence from a common point of departure, this continuity would be discernible in attenuated specimens.

In the ham, or the part of the frame brought into full play in the effort of leaping, is an important system of sinews; and another sinew, a double one, is that called 'the tendon', and others are those brought into play when a great effort of physical strength is required; that is to say, the epitonos or back-stay and the shoulder-sinews. Other sinews, devoid of specific designation, are situated in the region of the flexures of the bones; for all the bones that are attached to one another are bound together by sinews, and a great quantity of sinews are placed in the neighbourhood of all the bones. Only, by the way, in the head there is no sinew; but the head is held together by the sutures of the bones.

Sinew is fissile lengthwise, but crosswise it is not easily broken, but admits of a considerable amount of hard tension. In connexion with sinews a liquid mucus is developed, white and glutinous, and the organ, in fact, is sustained by it and appears to be substantially composed of it. Now, vein may be submitted to the actual cautery, but sinew, when submitted to such action, shrivels up altogether; and, if sinews be cut asunder, the severed parts will not again cohere. A feeling of numbness is incidental only to parts of the frame where sinew is situated.

There is a very extensive system of sinews connected severally with the feet, the hands, the ribs, the shoulder-blades, the neck, and the arms.

All animals supplied with blood are furnished with sinews; but in the case of animals that have no flexures to their limbs, but are, in fact, destitute of either feet or hands, the sinews are fine and inconspicuous; and so, as might have been anticipated, the sinews in the fish are chiefly discernible in connexion with the fin.

Part 6

The ines (or fibrous connective tissue) are a something intermediate between sinew and vein. Some of them are supplied with fluid, the lymph; and they pass from sinew to vein and from vein to sinew. There is another kind of ines or fibre that is found in blood, but not in the blood of all animals alike. If this fibre be left in the blood, the blood will coagulate; if it be removed or extracted, the blood is found to be incapable of coagulation. While, however, this fibrous matter is found in the blood of the great majority of animals, it is not found in all. For instance, we fail to find it in the blood of the deer, the roe, the antelope, and some other animals; and, owing to this deficiency of the fibrous tissue, the blood of these animals does not coagulate to the extent observed in the blood of other animals. The blood of the deer coagulates to about the same extent as that of the hare: that is to the blood in either case coagulates, but not into a stiff or jelly-like substance, like the blood of ordinary animals, but only into a flaccid consistency like that of milk which is not subjected to the action of rennet. The blood of the antelope admits of a firmer consistency in coagulation; for in this respect it resembles, or only comes a little short of, the blood of sheep. Such are the properties of vein, sinew, and fibrous tissue.

Part 7

The bones in animals are all connected with one single bone, and are interconnected, like the veins, in one unbroken sequence; and there is no instance of a bone standing apart by itself. In all animals furnished with bones, the spine or backbone is the point of origin for the entire osseous system. The spine is composed of vertebrae, and it extends from the head down to the loins. The vertebrae are all perforated, and, above, the bony portion of the head is connected with the topmost vertebrae, and is designated the 'skull'. And the serrated lines on the skull are termed 'sutures'.

The skull is not formed alike in all animals. In some animals the skull consists of one single undivided bone, as in the case of the dog; in others it is composite in structure, as in man; and in the human species the suture is circular in the female, while in the male it is made up of three separate sutures, uniting above in three-corner fashion; and instances have been known of a man's skull being devoid of suture altogether. The skull is composed not of four bones, but of six; two of these are in the region of the ears, small in comparison with the other four. From the skull extend the jaws, constituted of bone. (Animals in general move the lower jaw; the river crocodile is the only animal that moves the upper one.) In the jaws is the tooth-system; and the teeth are constituted of bone, and are half-way perforated; and the bone in question is the only kind of bone which it is found impossible to grave with a graving tool.

On the upper part of the course of the backbone are the collar-bones and the ribs. The chest rests on ribs; and these ribs meet together, whereas the others do not; for no animal has bone in the region of the stomach. Then come the shoulder-bones, or blade-bones, and the arm-bones connected with these, and the bones in the hands connected with the bones of the arms. With animals that have forelegs, the osseous system of the foreleg resembles that of the arm in man.

Below the level of the backbone, after the haunch-bone, comes the hip-socket; then the leg-bones, those in the thighs and those in the shins, which are termed colenes or limb-bones, a part of which is the ankle, while a part of the same is the so-called 'plectrum' in those creatures that have an ankle; and connected with these bones are the bones in the feet.

Now, with all animals that are supplied with blood and furnished with feet, and are at the same time viviparous, the bones do not differ greatly one from another, but only in the way of relative hardness, softness, or magnitude. A further difference, by the way, is that in one and the same animal certain bones are supplied with marrow, while others are destitute of it. Some animals might on casual observation appear to have no marrow whatsoever in their bones: as is the case with the lion, owing to his having marrow only in small amount, poor and thin, and in very few bones; for marrow is found in his thigh and arm bones. The bones of the lion are exceptionally hard; so hard, in fact, that if they are rubbed hard against one another they emit sparks like flint-stones. The dolphin has bones, and not fish-spine.

Of the other animals supplied with blood, some differ but little, as is the case with birds; others have systems analogous, as fishes; for viviparous fishes, such as the cartilaginous species, are gristle-spined, while the ovipara have a spine which corresponds to the backbone in quadrupeds. This exceptional property has been observed in fishes, that in some of them there are found delicate spines scattered here and there throughout the fleshy parts. The serpent is similarly constructed to the fish; in other words, his backbone is spinous. With oviparous quadrupeds, the skeleton of the larger ones is more or less osseous; of the smaller ones, more or less spinous. But all sanguineous animals have a backbone of either one kind or other: that is, composed either of bone or of spine.

The other portions of the skeleton are found in some animals and not found in others, but the presence or the absence of this and that part carries with it, as a matter of course, the presence or the absence of the bones or the spines corresponding to this or that part. For animals that are destitute of arms and legs cannot be furnished with limb-bones: and in like manner with animals that have the same parts, but yet have them unlike in form; for in these animals the corresponding bones differ from one another in the way of relative excess or relative defect, or in the way of analogy taking the place of identity. So much for the osseous or spinous systems in animals.

Part 8

Gristle is of the same nature as bone, but differs from it in the way of relative excess or relative defect. And just like bone, cartilage also, if cut, does not grow again. In terrestrial viviparous sanguinea the gristle formations are unperforated, and there is no marrow in them as there is in bones; in the selachia, however--for, be it observed, they are gristle-spined--there is found in the case of the flat space in the region of the backbone, a gristle-like substance analogous to bone, and in this gristle-like substance there is a liquid resembling marrow. In viviparous animals furnished with feet, gristle formations are found in the region of the ears, in the nostrils, and around certain extremities of the bones.

Part 9

Furthermore, there are parts of other kinds, neither identical with, nor altogether diverse from, the parts above enumerated: such as nails, hooves, claws, and horns; and also, by the way, beaks, such as birds are furnished with-all in the several animals that are furnished therewithal. All these parts are flexible and fissile; but bone is neither flexible nor fissile, but frangible.

And the colours of horns and nails and claw and hoof follow the colour of the skin and the hair. For according as the skin of an animal is black, or white, or of medium hue, so are the horns, the claws, or the hooves, as the case may be, of hue to match. And it is the same with nails. The teeth, however, follow after the bones. Thus in black men, such as the Aethiopians and the like, the teeth and bones are white, but the nails are black, like the whole of the skin.

Horns in general are hollow at their point of attachment to the bone which juts out from the head inside the horn, but they have a solid portion at the tip, and they are simple and undivided in structure. In the case of the stag alone of all animals the horns are solid throughout, and ramify into branches (or antlers). And, whereas no other animal is known to shed its horns, the deer sheds its horns annually, unless it has been castrated; and with regard to the effects of castration in animals we shall have much to say hereafter. Horns attach rather to the skin than to the bone; which will account for the fact that there are found in Phrygia and elsewhere cattle that can move their horns as freely as their ears.

Of animals furnished with nails-and, by the way, all animals have nails that have toes, and toes that have feet, except the elephant; and the elephant has toes undivided and slightly articulated, but has no nails whatsoever--of animals furnished with nails, some are straight-nailed, like man; others are crooked nailed, as the lion among animals that walk, and the eagle among animals that fly.

Part 10

The following are the properties of hair and of parts analogous to hair, and of skin or hide. All viviparous animals furnished with feet have hair; all oviparous animals furnished with feet have horn-like tessellates; fishes, and fishes only, have scales-that is, such oviparous fishes as have the crumbling egg or roe. For of the lanky fishes, the conger has no such egg, nor the muraena, and the eel has no egg at all.

The hair differs in the way of thickness and fineness, and of length, according to the locality of the part in which it is found, and according to the quality of skin or hide on which it grows. For, as a general rule, the thicker the hide, the harder and the thicker is the hair; and the hair is inclined to grow in abundance and to a great length in localities of the bodies hollow and moist, if the localities be fitted for the growth of hair at all. The facts are similar in the case of animals whether coated with scales or with tessellates. With soft-haired animals the hair gets harder with good feeding, and with hard-haired or bristly animals it gets softer and scantier from the same cause. Hair differs in quality also according to the relative heat or warmth of the locality: just as the hair in man is hard in warm places and soft in cold ones. Again, straight hair is inclined to be soft, and curly hair to be bristly.

Part 11

Hair is naturally fissile, and in this respect it differs in degree in diverse animals. In some animals the hair goes on gradually hardening into bristle until it no longer resembles hair but spine, as in the case of the hedgehog. And in like manner with the nails; for in some animals the nail differs as regards solidity in no way from bone.

Of all animals man has the most delicate skin: that is, if we take into consideration his relative size. In the skin or hide of all animals there is a mucous liquid, scanty in some animals and plentiful in others, as, for instance, in the hide of the ox; for men manufacture glue out of it. (And, by the way, in some cases glue is manufactured from fishes also.) The skin, when cut, is in itself devoid of sensation; and this is especially the case with the skin on the head, owing to there being no flesh between it and the skull. And wherever the skin is quite by itself, if it be cut asunder, it does not grow together again, as is seen in the thin part of the jaw, in the prepuce, and the eyelid. In all animals the skin is one of the parts that extends continuous and unbroken, and it comes to a stop only where the natural ducts pour out their contents, and at the mouth and nails.

All sanguineous animals, then, have skin; but not all such animals have hair, save only under the circumstances described above. The hair changes its colour as animals grow old, and in man it turns white or grey. With animals, in general, the change takes place, but not very obviously, or not so obviously as in the case of the horse. Hair turns grey from the point backwards to the roots. But, in the majority of cases, grey hairs are white from the beginning; and this is a proof that greyness of hair does not, as some believe to be the case, imply withering or decrepitude, for no part is brought into existence in a withered or decrepit condition.

In the eruptive malady called the white-sickness all the hairs get grey; and instances have been known where the hair became grey while the patients were ill of the malady, whereas the grey hairs shed off and black ones replaced them on their recovery. (Hair is more apt to turn grey when it is kept covered than when exposed to the action of the outer air.) In men, the hair over the temples is the first to turn grey, and the hair in the front grows grey sooner than the hair at the back; and the hair on the pubes is the last to change colour.

Some hairs are congenital, others grow after the maturity of the animal; but this occurs in man only. The congenital hairs are on the head, the eyelids, and the eyebrows; of the later growths the hairs on the pubes are the first to come, then those under the armpits, and, thirdly, those on the chin; for, singularly enough, the regions where congenital growths and the subsequent growths are found are equal in number. The hair on the head grows scanty and sheds out to a greater extent and sooner than all the rest. But this remark applies only to hair in front; for no man ever gets bald at the back of his head. Smoothness on the top of the head is termed 'baldness', but smoothness on the eyebrows is denoted by a special term which means 'forehead-baldness'; and neither of these conditions of baldness supervenes in a man until he shall have come under the influence of sexual passion. For no boy ever gets bald, no woman, and no castrated man. In fact, if a man be castrated before reaching puberty, the later growths of hair never come at all; and, if the operation take place subsequently, the aftergrowths, and these only, shed off; or, rather, two of the growths shed off, but not that on the pubes.

Women do not grow hairs on the chin; except that a scanty beard grows on some women after the monthly courses have stopped; and similar phenomenon is observed at times in priestesses in Caria, but these cases are looked upon as portentous with regard to coming events. The other after-growths are found in women, but more scanty and sparse. Men and women are at times born constitutionally and congenitally incapable of the after-growths; and individuals that are destitute even of the growth upon the pubes are constitutionally impotent.

Hair as a rule grows more or less in length as the wearer grows in age; chiefly the hair on the head, then that in the beard, and fine hair grows longest of all. With some people as they grow old the eyebrows grow thicker, to such an extent that they have to be cut off; and this growth is owing to the fact that the eyebrows are situated at a conjuncture of bones, and these bones, as age comes on, draw apart and exude a gradual increase of moisture or rheum. The eyelashes do not grow in size, but they shed when the wearer comes first under the influence of sexual feelings, and shed all the quicker as this influence is the more powerful; and these are the last hairs to grow grey.

Hairs if plucked out before maturity grow again; but they do not grow again if plucked out afterwards. Every hair is supplied with a mucous moisture at its root, and immediately after being plucked out it can lift light articles if it touch them with this mucus.

Animals that admit of diversity of colour in the hair admit of a similar diversity to start with in the skin and in the cuticle of the tongue.

In some cases among men the upper lip and the chin is thickly covered with hair, and in other cases these parts are smooth and the cheeks are hairy; and, by the way, smooth-chinned men are less inclined than bearded men to baldness.

The hair is inclined to grow in certain diseases, especially in consumption, and in old age, and after death; and under these circumstances the hair hardens concomitantly with its growth, and the same duplicate phenomenon is observable in respect of the nails.

In the case of men of strong sexual passions the congenital hairs shed the sooner, while the hairs of the after-growths are the quicker to come. When men are afflicted with varicose veins they are less inclined to take on baldness; and if they be bald when they become thus afflicted, they have a tendency to get their hair again.

If a hair be cut, it does not grow at the point of section; but it gets longer by growing upward from below. In fishes the scales grow harder and thicker with age, and when the animal gets emaciated or is growing old the scales grow harder. In quadrupeds as they grow old, the hair in some and the wool in others gets deeper but scantier in amount: and the hooves or claws get larger in size; and the same is the case with the beaks of birds. The claws also increase in size, as do also the nails.

Part 12

With regard to winged animals, such as birds, no creature is liable to change of colour by reason of age, excepting the crane. The wings of this bird are ash-coloured at first, but as it grows old the wings get black. Again, owing to special climatic influences, as when unusual frost prevails, a change is sometimes observed to take place in birds whose plumage is of one uniform colour; thus, birds that have dusky or downright black plumage turn white or grey, as the raven, the sparrow, and the swallow; but no case has ever yet been known of a change of colour from white to black. (Further, most birds change the colour of their plumage at different seasons of the year, so much so that a man ignorant of their habits might be mistaken as to their identity.) Some animals change the colour of their hair with a change in their drinking-water, for in some countries the same species of animal is found white in one district and black in another. And in regard to the commerce of the sexes, water in many places is of such peculiar quality that rams, if they have intercourse with the female after drinking it, beget black lambs, as is the case with the water of the Psychrus (so-called from its coldness), a river in the district of Assyritis in the Chalcidic Peninsula, on the coast of Thrace; and in Antandria there are two rivers of which one makes the lambs white and the other black. The river Scamander also has the reputation of making lambs yellow, and that is the reason, they say, why Homer designates it the 'Yellow River.' Animals as a general rule have no hair on their internal surfaces, and, in regard to their extremities, they have hair on the upper, but not on the lower side.

The hare, or dasypod, is the only animal known to have hair inside its mouth and underneath its feet. Further, the so-called mousewhale instead of teeth has hairs in its mouth resembling pigs' bristles.

Hairs after being cut grow at the bottom but not at the top; if feathers be cut off, they grow neither at top nor bottom, but shed and fall out. Further, the bee's wing will not grow again after being plucked off, nor will the wing of any creature that has undivided wings. Neither will the sting grow again if the bee lose it, but the creature will die of the loss.

Part 13

In all sanguineous animals membranes are found. And membrane resembles a thin close-textured skin, but its qualities are different, as it admits neither of cleavage nor of extension. Membrane envelops each one of the bones and each one of the viscera, both in the larger and the smaller animals; though in the smaller animals the membranes are indiscernible from their extreme tenuity and minuteness. The largest of all the membranes are the two that surround the brain, and of these two the one that lines the bony skull is stronger and thicker than the one that envelops the brain; next in order of magnitude comes the membrane that encloses the heart. If membrane be bared and cut asunder it will not grow together again, and the bone thus stripped of its membrane mortifies.

Part 14

The omentum or caul, by the way, is membrane. All sanguineous animals are furnished with this organ; but in some animals the organ is supplied with fat, and in others it is devoid of it. The omentum has both its starting-point and its attachment, with ambidental vivipara, in the centre of the stomach, where the stomach has a kind of suture; in non-ambidental vivipara it has its starting-point and attachment in the chief of the ruminating stomachs.

Part 15

The bladder also is of the nature of membrane, but of membrane peculiar in kind, for it is extensile. The organ is not common to all animals, but, while it is found in all the vivipara, the tortoise is the only oviparous animal that is furnished therewithal. The bladder, like ordinary membrane, if cut asunder will not grow together again, unless the section be just at the commencement of the urethra: except indeed in very rare cases, for instances of healing have been known to occur. After death, the organ passes no liquid excretion; but in life, in addition to the normal liquid excretion, it passes at times dry excretion also, which turns into stones in the case of sufferers from that malady. Indeed, instances have been known of concretions in the bladder so shaped as closely to resemble cockleshells.

Such are the properties, then, of vein, sinew and skin, of fibre and membrane, of hair, nail, claw and hoof, of horns, of teeth, of beak, of gristle, of bones, and of parts that are analogous to any of the parts here enumerated.

Part 16

Flesh, and that which is by nature akin to it in sanguineous animals, is in all cases situated in between the skin and the bone, or the substance analogous to bone; for just as spine is a counterpart of bone, so is the flesh-like substance of animals that are constructed a spinous system the counterpart of the flesh of animals constructed on an osseous one.

Flesh can be divided asunder in any direction, not lengthwise only as is the case with sinew and vein. When animals are subjected to emaciation the flesh disappears, and the creatures become a mass of veins and fibres; when they are over fed, fat takes the place of flesh. Where the flesh is abundant in an animal, its veins are somewhat small and the blood abnormally red; the viscera also and the stomach are diminutive; whereas with animals whose veins are large the blood is somewhat black, the viscera and the stomach are large, and the flesh is somewhat scanty. And animals with small stomachs are disposed to take on flesh.

Part 17

Again, fat and suet differ from one another. Suet is frangible in all directions and congeals if subjected to extreme cold, whereas fat can melt but cannot freeze or congeal; and soups made of the flesh of animals supplied with fat do not congeal or coagulate, as is found with horse-flesh and pork; but soups made from the flesh of animals supplied with suet do coagulate, as is seen with mutton and goat's flesh. Further, fat and suet differ as to their localities: for fat is found between the skin and flesh, but suet is found only at the limit of the fleshy parts. Also, in animals supplied with fat the omentum or caul is supplied with fat, and it is supplied with suet in animals supplied with suet. Moreover, ambidental animals are supplied with fat, and non-ambidentals with suet.

Of the viscera the liver in some animals becomes fatty, as, among fishes, is the case with the selachia, by the melting of whose livers an oil is manufactured. These cartilaginous fish themselves have no free fat at all in connexion with the flesh or with the stomach. The suet in fish is fatty, and does not solidify or congeal. All animals are furnished with fat, either intermingled with their flesh, or apart. Such as have no free or separate fat are less fat than others in stomach and omentum, as the eel; for it has only a scanty supply of suet about the omentum. Most animals take on fat in the belly, especially such animals as are little in motion.

The brains of animals supplied with fat are oily, as in the pig; of animals supplied with suet, parched and dry. But it is about the kidneys more than any other viscera that animals are inclined to take on fat; and the right kidney is always less supplied with fat than the left kidney, and, be the two kidneys ever so fat, there is always a space devoid of fat in between the two. Animals supplied with suet are specially apt to have it about the kidneys, and especially the sheep; for this animal is apt to die from its kidneys being entirely enveloped. Fat or suet about the kidney is superinduced by overfeeding, as is found at Leontini in Sicily; and consequently in this district they defer driving out sheep to pasture until the day is well on, with the view of limiting their food by curtailment of the hours of pasture.

Part 18

The part around the pupil of the eye is fatty in all animals, and this part resembles suet in all animals that possess such a part and that are not furnished with hard eyes.

Fat animals, whether male or female, are more or less unfitted for breeding purposes. Animals are disposed to take on fat more when old than when young, and especially when they have attained their full breadth and their full length and are beginning to grow depthways.

Part 19

And now to proceed to the consideration of the blood. In sanguineous animals blood is the most universal and the most indispensable part; and it is not an acquired or adventitious part, but it is a consubstantial part of all animals that are not corrupt or moribund. All blood is contained in a vascular system, to wit, the veins, and is found nowhere else, excepting in the heart. Blood is not sensitive to touch in any animal, any more than the excretions of the stomach; and the case is similar with the brain and the marrow. When flesh is lacerated, blood exudes, if the animal be alive and unless the flesh be gangrened. Blood in a healthy condition is naturally sweet to the taste, and red in colour, blood that deteriorates from natural decay or from disease more or less black. Blood at its best, before it undergoes deterioration from either natural decay or from disease, is neither very thick nor very thin. In the living animal it is always liquid and warm, but, on issuing from the body, it coagulates in all cases except in the case of the deer, the roe, and the like animals; for, as a general rule, blood coagulates unless the fibres be extracted. Bull's blood is the quickest to coagulate.

Animals that are internally and externally viviparous are more abundantly supplied with blood than the sanguineous ovipara. Animals that are in good condition, either from natural causes or from their health having been attended to, have the blood neither too abundant-as creatures just after drinking have the liquid inside them in abundance-nor again very scanty, as is the case with animals when exceedingly fat. For animals in this condition have pure blood, but very little of it, and the fatter an animal gets the less becomes its supply of blood; for whatsoever is fat is destitute of blood.

A fat substance is incorruptible, but blood and all things containing it corrupt rapidly, and this property characterizes especially all parts connected with the bones. Blood is finest and purest in man; and thickest and blackest in the bull and the ass, of all vivipara. In the lower and the higher parts of the body blood is thicker and blacker than in the central parts.

Blood beats or palpitates in the veins of all animals alike all over their bodies, and blood is the only liquid that permeates the entire frames of living animals, without exception and at all times, as long as life lasts. Blood is developed first of all in the heart of animals before the body is differentiated as a whole. If blood be removed or if it escape in any considerable quantity, animals fall into a faint or swoon; if it be removed or if it escape in an exceedingly large quantity they die. If the blood get exceedingly liquid, animals fall sick; for the blood then turns into something like ichor, or a liquid so thin that it at times has been known to exude through the pores like sweat. In some cases blood, when issuing from the veins, does not coagulate at all, or only here and there. Whilst animals are sleeping the blood is less abundantly supplied near the exterior surfaces, so that, if the sleeping creature be pricked with a pin, the blood does not issue as copiously as it would if the creature were awake. Blood is developed out of ichor by coction, and fat in like manner out of blood. If the blood get diseased, hemorrhoids may ensue in the nostril or at the anus, or the veins may become varicose. Blood, if it corrupt in the body, has a tendency to turn into pus, and pus may turn into a solid concretion.

Blood in the female differs from that in the male, for, supposing the male and female to be on a par as regards age and general health, the blood in the female is thicker and blacker than in the male; and with the female there is a comparative superabundance of it in the interior. Of all female animals the female in man is the most richly supplied with blood, and of all female animals the menstruous discharges are the most copious in woman. The blood of these discharges under disease turns into flux. Apart from the menstrual discharges, the female in the human species is less subject to diseases of the blood than the male.

Women are seldom afflicted with varicose veins, with hemorrhoids, or with bleeding at the nose, and, if any of these maladies supervene, the menses are imperfectly discharged.

Blood differs in quantity and appearance according to age; in very young animals it resembles ichor and is abundant, in the old it is thick and black and scarce, and in middle-aged animals its qualities are intermediate. In old animals the blood coagulates rapidly, even blood at the surface of the body; but this is not the case with young animals. Ichor is, in fact, nothing else but unconcocted blood: either blood that has not yet been concocted, or that has become fluid again.

Part 20

We now proceed to discuss the properties of marrow; for this is one of the liquids found in certain sanguineous animals. All the natural liquids of the body are contained in vessels: as blood in veins, marrow in bones other moistures in membranous structures of the skin

In young animals the marrow is exceedingly sanguineous, but, as animals grow old, it becomes fatty in animals supplied with fat, and suet-like in animals with suet. All bones, however, are not supplied with marrow, but only the hollow ones, and not all of these. For of the bones in the lion some contain no marrow at all, and some are only scantily supplied therewith; and that accounts, as was previously observed, for the statement made by certain writers that the lion is marrowless. In the bones of pigs it is found in small quantities; and in the bones of certain animals of this species it is not found at all.

These liquids, then, are nearly always congenital in animals, but milk and sperm come at a later time. Of these latter, that which, whensoever it is present, is secreted in all cases ready-made, is the milk; sperm, on the other hand, is not secreted out in all cases, but in some only, as in the case of what are designated thori in fishes.

Whatever animals have milk, have it in their breasts. All animals have breasts that are internally and externally viviparous, as for instance all animals that have hair, as man and the horse; and the cetaceans, as the dolphin, the porpoise, and the whale-for these animals have breasts and are supplied with milk. Animals that are oviparous or only externally viviparous have neither breasts nor milk, as the fish and the bird.

All milk is composed of a watery serum called 'whey', and a consistent substance called curd (or cheese); and the thicker the milk, the more abundant the curd. The milk, then, of non-ambidentals coagulates, and that is why cheese is made of the milk of such animals under domestication; but the milk of ambidentals does not coagulate, nor their fat either, and the milk is thin and sweet. Now the camel's milk is the thinnest, and that of the human species next after it, and that of the ass next again, but cow's milk is the thickest. Milk does not coagulate under the influence of cold, but rather runs to whey; but under the influence of heat it coagulates and thickens. As a general rule milk only comes to animals in pregnancy. When the animal is pregnant milk is found, but for a while it is unfit for use, and then after an interval of usefulness it becomes unfit for use again. In the case of female animals not pregnant a small quantity of milk has been procured by the employment of special food, and cases have been actually known where women advanced in years on being submitted to the process of milking have produced milk, and in some cases have produced it in sufficient quantities to enable them to suckle an infant.

The people that live on and about Mount Oeta take such she-goats as decline the male and rub their udders hard with nettles to cause an irritation amounting to pain; hereupon they milk the animals, procuring at first a liquid resembling blood, then a liquid mixed with purulent matter, and eventually milk, as freely as from females submitting to the male.

As a general rule, milk is not found in the male of man or of any other animal, though from time to time it has been found in a male; for instance, once in Lemnos a he-goat was milked by its dugs (for it has, by the way, two dugs close to the penis), and was milked to such effect that cheese was made of the produce, and the same phenomenon was repeated in a male of its own begetting. Such occurrences, however, are regarded as supernatural and fraught with omen as to futurity, and in point of fact when the Lemnian owner of the animal inquired of the oracle, the god informed him that the portent foreshadowed the acquisition of a fortune. With some men, after puberty, milk can be produced by squeezing the breasts; cases have been known where on their being subjected to a prolonged milking process a considerable quantity of milk has been educed.

In milk there is a fatty element, which in clotted milk gets to resemble oil. Goat's milk is mixed with sheep's milk in Sicily, and wherever sheep's milk is abundant. The best milk for clotting is not only that where the cheese is most abundant, but that also where the cheese is driest.

Now some animals produce not only enough milk to rear their young, but a superfluous amount for general use, for cheese-making and for storage. This is especially the case with the sheep and the goat, and next in degree with the cow. Mare's milk, by the way, and milk of the she-ass are mixed in with Phrygian cheese. And there is more cheese in cow's milk than in goat's milk; for graziers tell us that from nine gallons of goat's milk they can get nineteen cheeses at an obol apiece, and from the same amount of cow's milk, thirty. Other animals give only enough of milk to rear their young withal, and no superfluous amount and none fitted for cheese-making, as is the case with all animals that have more than two breasts or dugs; for with none of such animals is milk produced in superabundance or used for the manufacture of cheese.

The juice of the fig and rennet are employed to curdle milk. The fig-juice is first squeezed out into wool; the wool is then washed and rinsed, and the rinsing put into a little milk, and if this be mixed with other milk it curdles Rennet is a kind of milk, for it is found in the stomach of the animal while it is yet suckling.

Part 21

Rennet then consists of milk with an admixture of fire, which comes from the natural heat of the animal, as the milk is concocted. All ruminating animals produce rennet, and, of ambidentals, the hare. Rennet improves in quality the longer it is kept; and cow's rennet, after being kept a good while, and also hare's rennet, is good for diarrhoea, and the best of all rennet is that of the young deer.

In milk-producing animals the comparative amount of the yield varies with the size of the animal and the diversities of pasturage. For instance, there are in Phasis small cattle that in all cases give a copious supply of milk, and the large cows in Epirus yield each one daily some nine gallons of milk, and half of this from each pair of teats, and the milker has to stand erect, stooping forward a little, as otherwise, if he were seated, he would be unable to reach up to the teats. But, with the exception of the ass, all the quadrupeds in Epirus are of large size, and relatively, the cattle and the dogs are the largest. Now large animals require abundant pasture, and this country supplies just such pasturage, and also supplies diverse pasture grounds to suit the diverse seasons of the year. The cattle are particularly large, and likewise the sheep of the so-called Pyrrhic breed, the name being given in honour of King Pyrrhus.

Some pasture quenches milk, as Median grass or lucerne, and that especially in ruminants; other feeding renders it copious, as cytisus and vetch; only, by the way, cytisus in flower is not recommended, as it has burning properties, and vetch is not good for pregnant kine, as it causes increased difficulty in parturition. However, beasts that have access to good feeding, as they are benefited thereby in regard to pregnancy, so also being well nourished produce milk in plenty. Some of the leguminous plants bring milk in abundance, as for instance, a large feed of beans with the ewe, the common she-goat, the cow, and the small she-goat; for this feeding makes them drop their udders. And, by the way, the pointing of the udder to the ground before parturition is a sign of there being plenty of milk coming.

Milk remains for a long time in the female, if she be kept from the male and be properly fed, and, of quadrupeds, this is especially true of the ewe; for the ewe can be milked for eight months. As a general rule, ruminating animals give milk in abundance, and milk fitted for cheese manufacture. In the neighbourhood of Torone cows run dry for a few days before calving, and have milk all the rest of the time. In women, milk of a livid colour is better than white for nursing purposes; and swarthy women give healthier milk than fair ones. Milk that is richest in cheese is the most nutritious, but milk with a scanty supply of cheese is the more wholesome for children.

Part 22

All sanguineous animals eject sperm. As to what, and how, it contributes to generation, these questions will be discussed in another treatise. Taking the size of his body into account, man emits more sperm than any other animal. In hairy-coated animals the sperm is sticky, but in other animals it is not so. It is white in all cases, and Herodotus is under a misapprehension when he states that the Ethiopians eject black sperm.

Sperm issues from the body white and consistent, if it be healthy, and after quitting the body becomes thin and black. In frosty weather it does not coagulate, but gets exceedingly thin and watery both in colour and consistency; but it coagulates and thickens under the influence of heat. If it be long in the womb before issuing out, it comes more than usually thick; and sometimes it comes out dry and compact. Sperm capable of impregnating or of fructification sinks in water; sperm incapable of producing that result dissolves away. But there is no truth in what Ctesias has written about the sperm of the elephant.

Book IV

Part 1

WE have now treated, in regard to blooded animals of the parts they have in common and of the parts peculiar to this genus or that, and of the parts both composite and simple, whether without or within. We now proceed to treat of animals devoid of blood. These animals are divided into several genera.

One genus consists of so-called 'molluscs'; and by the term 'mollusc' we mean an animal that, being devoid of blood, has its flesh-like substance outside, and any hard structure it may happen to have, inside-in this respect resembling the red-blooded animals, such as the genus of the cuttle-fish.

Another genus is that of the malacostraca. These are animals that have their hard structure outside, and their soft or fleshlike substance inside, and the hard substance belonging to them has to be crushed rather than shattered; and to this genus belongs the crawfish and the crab.

A third genus is that of the ostracoderms or 'testaceans'. These are animals that have their hard substance outside and their flesh-like substance within, and their hard substance can be shattered but not crushed; and to this genus belong the snail and the oyster.

The fourth genus is that of insects; and this genus comprehends numerous and dissimilar species. Insects are creatures that, as the name implies, have nicks either on the belly or on the back, or on both belly and back, and have no one part distinctly osseous and no one part distinctly fleshy, but are throughout a something intermediate between bone and flesh; that is to say, their body is hard all through, inside and outside. Some insects are wingless, such as the iulus and the centipede; some are winged, as the bee, the cockchafer, and the wasp; and the same kind is in some cases both winged and wingless, as the ant and the glow-worm.

In molluscs the external parts are as follows: in the first place, the so-called feet; secondly, and attached to these, the head; thirdly, the mantle-sac, containing the internal parts, and incorrectly designated by some writers the head; and, fourthly, fins round about the sac. (See diagram.) In all molluscs the head is found to be between the feet and the belly. All molluscs are furnished with eight feet, and in all cases these feet are severally furnished with a double row of suckers, with the exception of one single species of poulpe or octopus. The sepia, the small calamari and the large calamary have an exceptional organ in a pair of long arms or tentacles, having at their extremities a portion rendered rough by the presence of two rows of suckers; and with these arms or tentacles they apprehend their food and draw it into their mouths, and in stormy weather they cling by them to a rock and sway about in the rough water like ships lying at anchor. They swim by the aid of the fins that they have about the sac. In all cases their feet are furnished with suckers.

The octopus, by the way, uses his feelers either as feet or hands; with the two which stand over his mouth he draws in food, and the last of his feelers he employs in the act of copulation; and this last one, by the way, is extremely sharp, is exceptional as being of a whitish colour, and at its extremity is bifurcate; that is to say, it has an additional something on the rachis, and by rachis is meant the smooth surface or edge of the arm on the far side from the suckers. (See diagram.)

In front of the sac and over the feelers they have a hollow tube, by means of which they discharge any sea-water that they may have taken into the sac of the body in the act of receiving food by the mouth. They can shift the tube from side to side, and by means of it they discharge the black liquid peculiar to the animal.

Stretching out its feet, it swims obliquely in the direction of the so-called head, and by this mode of swimming it can see in front, for its eyes are at the top, and in this attitude it has its mouth at the rear. The 'head', while the creature is alive, is hard, and looks as though it were inflated. It apprehends and retains objects by means of the under-surface of its arms, and the membrane in between its feet is kept at full tension; if the animal get on to the sand it can no longer retain its hold.

There is a difference between the octopus and the other molluscs above mentioned: the body of the octopus is small, and his feet are long, whereas in the others the body is large and the feet short; so short, in fact, that they cannot walk on them. Compared with one another, the teuthis, or calamari, is long-shaped, and the sepia flat-shaped; and of the calamaries the so-called teuthus is much bigger than the teuthis; for teuthi have been found as much as five ells long. Some sepiae attain a length of two ells, and the feelers of the octopus are sometimes as long, or even longer. The species teuthus is not a numerous one; the teuthus differs from the teuthis in shape; that is, the sharp extremity of the teuthus is broader than that of the other, and, further, the encircling fin goes all round the trunk, whereas it is in part lacking in the teuthis; both animals are pelagic.

In all cases the head comes after the feet, in the middle of the feet that are called arms or feelers. There is here situated a mouth, and two teeth in the mouth; and above these two large eyes, and betwixt the eyes a small cartilage enclosing a small brain; and within the mouth it has a minute organ of a fleshy nature, and this it uses as a tongue, for no other tongue does it possess. Next after this, on the outside, is what looks like a sac; the flesh of which it is made is divisible, not in long straight strips, but in annular flakes; and all molluscs have a cuticle around this flesh. Next after or at the back of the mouth comes a long and narrow esophagus, and close after that a crop or craw, large and spherical, like that of a bird; then comes the stomach, like the fourth stomach in ruminants; and the shape of it resembles the spiral convolution in the trumpet-shell; from the stomach there goes back again, in the direction of the mouth, thin gut, and the gut is thicker than the esophagus. (See diagram.)

Molluscs have no viscera, but they have what is called a mytis, and on it a vessel containing a thick black juice; in the sepia or cuttle-fish this vessel is the largest, and this juice is most abundant. All molluscs, when frightened, discharge such a juice, but the discharge is most copious in the cuttle-fish. The mytis, then, is situated under the mouth, and the esophagus runs through it; and down below at the point to which the gut extends is the vesicle of the black juice, and the animal has the vesicle and the gut enveloped in one and the same membrane, and by the same membrane, same orifice discharges both the black juice and the residuum. The animals have also certain hair-like or furry growths in their bodies.

In the sepia, the teuthis, and the teuthus the hard parts are within, towards the back of the body; those parts are called in one the sepium, and in the other the 'sword'. They differ from one another, for the sepium in the cuttle-fish and teuthus is hard and flat, being a substance intermediate between bone and fishbone, with (in part) a crumbling, spongy texture, but in the teuthis the part is thin and somewhat gristly. These parts differ from one another in shape, as do also the bodies of the animals. The octopus has nothing hard of this kind in its interior, but it has a gristly substance round the head, which, if the animal grows old, becomes hard.

The females differ from the males. The males have a duct in under the esophagus, extending from the mantle-cavity to the lower portion of the sac, and there is an organ to which it attaches, resembling a breast; (see diagram) in the female there are two of these organs, situated higher up; (see diagram) with both sexes there are underneath these organs certain red formations. The egg of the octopus is single, uneven on its surface, and of large size; the fluid substance within is all uniform in colour, smooth, and in colour white; the size of the egg is so great as to fill a vessel larger than the creature's head. The sepia has two sacs, and inside them a number of eggs, like in appearance to white hailstones. For the disposition of these parts, I must refer to my anatomical diagrams.

The males of all these animals differ from the females, and the difference between the sexes is most marked in the sepia; for the back of the trunk, which is blacker than the belly, is rougher in the male than in the female, and in the male the back is striped, and the rump is more sharply pointed.

There are several species of the octopus. One keeps close to the surface, and is the largest of them all, and near the shore the size is larger than in deep water; and there are others, small, variegated in colour, which are not articles of food. There are two others, one called the heledone, which differs from its congeners in the length of its legs and in having one row of suckers-all the rest of the molluscs having two,-the other nicknamed variously the bolitaina or the 'onion,' and the ozolis or the 'stinkard'.

There are two others found in shells resembling those of the testaceans. One of them is nicknamed by some persons the nautilus or the pontilus, or by others the 'polypus' egg'; and the shell of this creature is something like a separate valve of a deep scallop-shell. This polypus lives very often near to the shore, and is apt to be thrown up high and dry on the beach; under these circumstances it is found with its shell detached, and dies by and by on dry land. These polypods are small, and are shaped, as regards the form of their bodies, like the bolbidia. There is another polypus that is placed within a shell like a snail; it never comes out of the shell, but lives inside the shell like the snail, and from time to time protrudes its feelers.

So much for molluscs.

Part 2

With regard to the Malacostraca or crustaceans, one species is that of the crawfish, and a second, resembling the first, is that of the lobster; the lobster differing from the crawfish in having claws, and in a few other respects as well. Another species is that of the carid, and another is that of the crab, and there are many kinds both of carid and of crab.

Of carids there are the so-called cyphae, or 'hunch-backs', the crangons, or squillae, and the little kind, or shrimps, and the little kind do not develop into a larger kind.

Of the crab, the varieties are indefinite and incalculable. The largest of all crabs is one nicknamed Maia, a second variety is the pagarus and the crab of Heracleotis, and a third variety is the fresh-water crab; the other varieties are smaller in size and destitute of special designations. In the neighbourhood of Phoenice there are found on the beach certain crabs that are nicknamed the 'horsemen', from their running with such speed that it is difficult to overtake them; these crabs, when opened, are usually found empty, and this emptiness may be put down to insufficiency of nutriment. (There is another variety, small like the crab, but resembling in shape the lobster.) All these animals, as has been stated, have their hard and shelly part outside, where the skin is in other animals, and the fleshy part inside; and the belly is more or less provided with lamellae, or little flaps, and the female here deposits her spawn.

The crawfishes have five feet on either side, including the claws at the end; and in like manner the crabs have ten feet in all, including the claws. Of the carids, the hunch-backed, or prawns, have five feet on either side, which are sharp-pointed-those towards the head; and five others on either side in the region of the belly, with their extremities flat; they are devoid of flaps on the underside such as the crawfish has, but on the back they resemble the crawfish. (See diagram.)It is very different with the crangon, or squilla; it has four front legs on either side, then three thin ones close behind on either side, and the rest of the body is for the most part devoid of feet. (See diagram.) Of all these animals the feet bend out obliquely, as is the case with insects; and the claws, where claws are found, turn inwards. The crawfish has a tail, and five fins on it; and the round-backed carid has a tail and four fins; the squilla also has fins at the tail on either side. In the case of both the hump-backed carid and the squilla the middle art of the tail is spinous: only that in the squilla the part is flattened and in the carid it is sharp-pointed. Of all animals of this genus the crab is the only one devoid of a rump; and, while the body of the carid and the crawfish is elongated, that of the crab is rotund.

In the crawfish the male differs from the female: in the female the first foot is bifurcate, in the male it is undivided; the belly-fins in the female are large and overlapping on the neck, while in the male they are smaller and do not overlap; and, further, on the last feet of the male there are spur-like projections, large and sharp, which projections in the female are small and smooth. Both male and female have two antennae in front of the eyes, large and rough, and other antennae underneath, small and smooth. The eyes of all these creatures are hard and beady, and can move either to the inner or to the outer side. The eyes of most crabs have a similar facility of movement, or rather, in the crab this facility is developed in a higher degree. (See diagram.)

The lobster is all over grey-coloured, with a mottling of black. Its under or hinder feet, up to the big feet or claws, are eight in number; then come the big feet, far larger and flatter at the tips than the same organs in the crawfish; and these big feet or claws are exceptional in their structure, for the right claw has the extreme flat surface long and thin, while the left claw has the corresponding surface thick and round. Each of the two claws, divided at the end like a pair of jaws, has both below and above a set of teeth: only that in the right claw they are all small and saw-shaped, while in the left claw those at the apex are saw-shaped and those within are molar-shaped, these latter being, in the under part of the cleft claw, four teeth close together, and in the upper part three teeth, not close together. Both right and left claws have the upper part mobile, and bring it to bear against the lower one, and both are curved like bandy-legs, being thereby adapted for apprehension and constriction. Above the two large claws come two others, covered with hair, a little underneath the mouth; and underneath these the gill-like formations in the region of the mouth, hairy and numerous. These organs the animal keeps in perpetual motion; and the two hairy feet it bends and draws in towards its mouth. The feet near the mouth are furnished also with delicate outgrowing appendages. Like the crawfish, the lobster has two teeth, or mandibles, and above these teeth are its antennae, long, but shorter and finer by far than those of the crawfish, and then four other antennae similar in shape, but shorter and finer than the others. Over these antennae come the eyes, small and short, not large like the eyes of the crawfish. Over the eyes is a peaky rough projection like a forehead, larger than the same part in the crawfish; in fact, the frontal part is more pointed and the thorax is much broader in the lobster than in the crawfish, and the body in general is smoother and more full of flesh. Of the

eight feet, four are bifurcate at the extremities, and four are undivided. The region of the so-
called neck is outwardly divided into five divisions, and sixthly comes the flattened portion at the
end, and this portion has five flaps, or tail-fins; and the inner or under parts, into which the
female drops her spawn, are four in number and hairy, and on each of the aforesaid parts is a
spine turned outwards, short and straight. The body in general and the region of the thorax in
particular are smooth, not rough as in the crawfish; but on the large claws the outer portion has
larger spines. There is no apparent difference between the male and female, for they both have
one claw, whichever it may be, larger than the other, and neither male nor female is ever found
with both claws of the same size.

All crustaceans take in water close by the mouth. The crab discharges it, closing up, as it does so,
a small portion of the same, and the crawfish discharges it by way of the gills; and, by the way,
the gill-shaped organs in the crawfish are very numerous.

The following properties are common to all crustaceans: they have in all cases two teeth, or
mandibles (for the front teeth in the crawfish are two in number), and in all cases there is in the
mouth a small fleshy structure serving for a tongue; and the stomach is close to the mouth, only
that the crawfish has a little esophagus in front of the stomach, and there is a straight gut attached
to it. This gut, in the crawfish and its congeners, and in the carids, extends in a straight line to the
tail, and terminates where the animal discharges the residuum, and where the female deposits her
spawn; in the crab it terminates where the flap is situated, and in the centre of the flap. (And by
the way, in all these animals the spawn is deposited outside.) Further, the female has the place
for the spawn running along the gut. And, again, all these animals have, more or less, an organ
termed the 'mytis', or 'poppyjuice'.

We must now proceed to review their several differentiae. The crawfish then, as has been said,
has two teeth, large and hollow, in which is contained a juice resembling the mytis, and in
between the teeth is a fleshy substance, shaped like a tongue. After the mouth comes a short
esophagus, and then a membranous stomach attached to the esophagus, and at the orifice Of the
stomach are three teeth, two facing one another and a third standing by itself underneath.
Coming off at a bend from the stomach is a gut, simple and of equal thickness throughout the
entire length of the body until it reaches the anal vent.

These are all common properties of the crawfish, the carid, and the crab; for the crab, be it remembered, has two teeth.

Again, the crawfish has a duct attached all the way from the chest to the anal vent; and this duct is connected with the ovary in the female, and with the seminal ducts in the male. This passage is attached to the concave surface of the flesh in such a way that the flesh is in betwixt the duct and the gut; for the gut is related to the convexity and this duct to the concavity, pretty much as is observed in quadrupeds. And the duct is identical in both the sexes; that is to say, the duct in both is thin and white, and charged with a sallow-coloured moisture, and is attached to the chest.

(The following are the properties of the egg and of the convolutes in the carid.)

The male, by the way, differs from the female in regard to its flesh, in having in connexion with the chest two separate and distinct white substances, resembling in colour and conformation the tentacles of the cuttle-fish, and they are convoluted like the 'poppy' or quasi-liver of the trumpet-shell. These organs have their starting-point in 'cotyledons' or papillae, which are situated under the hindmost feet; and hereabouts the flesh is red and blood-coloured, but is slippery to the touch and in so far unlike flesh. Off from the convolute organ at the chest branches off another coil about as thick as ordinary twine; and underneath there are two granular seminal bodies in juxta-position with the gut. These are the organs of the male. The female has red-coloured eggs, which are adjacent to the stomach and to each side of the gut all along to the fleshy parts, being enveloped in a thin membrane.

Such are the parts, internal and external, of the carid.

Part 3

The inner organs of sanguineous animals happen to have specific designations; for these animals have in all cases the inner viscera, but this is not the case with the bloodless animals, but what they have in common with red-blooded animals is the stomach, the esophagus, and the gut.

With regard to the crab, it has already been stated that it has claws and feet, and their position has been set forth; furthermore, for the most part they have the right claw bigger and stronger than the left. It has also been stated' that in general the eyes of the crab look sideways. Further, the trunk of the crab's body is single and undivided, including its head and any other part it may possess. Some crabs have eyes placed sideways on the upper part, immediately under the back, and standing a long way apart, and some have their eyes in the centre and close together, like the crabs of Heracleotis and the so-called 'grannies'. The mouth lies underneath the eyes, and inside it there are two teeth, as is the case with the crawfish, only that in the crab the teeth are not rounded but long; and over the teeth are two lids, and in betwixt them are structures such as the crawfish has besides its teeth. The crab takes in water near by the mouth, using the lids as a check to the inflow, and discharges the water by two passages above the mouth, closing by means of the lids the way by which it entered; and the two passage-ways are underneath the eyes. When it has taken in water it closes its mouth by means of both lids, and ejects the water in the way above described. Next after the teeth comes the esophagus, very short, so short in fact that the stomach seems to come straightway after the mouth. Next after the esophagus comes the stomach, two-horned, to the centre of which is attached a simple and delicate gut; and the gut terminates outwards, at the operculum, as has been previously stated. (The crab has the parts in between the lids in the neighbourhood of the teeth similar to the same parts in the crawfish.) Inside the trunk is a sallow juice and some few little bodies, long and white, and others spotted red. The male differs from the female in size and breadth, and in respect of the ventral flap; for this is larger in the female than in the male, and stands out further from the trunk, and is more hairy (as is the case also with the female in the crawfish).

So much, then, for the organs of the malacostraca or crustacea.

Part 4

With the ostracoderma, or testaceans, such as the land-snails and the sea-snails, and all the 'oysters' so-called, and also with the sea-urchin genus, the fleshy part, in such as have flesh, is similarly situated to the fleshy part in the crustaceans; in other words, it is inside the animal, and the shell is outside, and there is no hard substance in the interior. As compared with one another the testaceans present many diversities both in regard to their shells and to the flesh within. Some of them have no flesh at all, as the sea-urchin; others have flesh, but it is inside and wholly hidden, except the head, as in the land-snails, and the so-called cocalia, and, among pelagic animals, in the purple murex, the ceryx or trumpet-shell, the sea-snail, and the spiral-shaped testaceans in general. Of the rest, some are bivalved and some univalved; and by 'bivalves' I mean such as are enclosed within two shells, and by 'univalved' such as are enclosed within a single shell, and in these last the fleshy part is exposed, as in the case of the limpet. Of the bivalves, some can open out, like the scallop and the mussel; for all such shells are grown together on one side and are separate on the other, so as to open and shut. Other bivalves are closed on both sides alike, like the solen or razor-fish. Some testaceans there are, that are entirely enveloped in shell and expose no portion of their flesh outside, as the tethya or ascidians.

Again, in regard to the shells themselves, the testaceans present differences when compared with one another. Some are smooth-shelled, like the solen, the mussel, and some clams, viz. those that are nicknamed 'milkshells', while others are rough-shelled, such as the pool-oyster or edible oyster, the pinna, and certain species of cockles, and the trumpet shells; and of these some are ribbed, such as the scallop and a certain kind of clam or cockle, and some are devoid of ribs, as the pinna and another species of clam. Testaceans also differ from one another in regard to the thickness or thinness of their shell, both as regards the shell in its entirety and as regards specific parts of the shell, for instance, the lips; for some have thin-lipped shells, like the mussel, and others have thick-lipped shells, like the oyster. A property common to the above mentioned, and, in fact, to all testaceans, is the smoothness of their shells inside. Some also are capable of motion, like the scallop, and indeed some aver that scallops can actually fly, owing to the circumstance that they often jump right out of the apparatus by means of which they are caught; others are incapable of motion and are attached fast to some external object, as is the case with the pinna. All the spiral-shaped testaceans can move and creep, and even the limpet relaxes its hold to go in quest of food. In the case of the univalves and the bivalves, the fleshy substance adheres to the shell so tenaciously that it can only be removed by an effort; in the case of the stromboids, it is more loosely attached. And a peculiarity of all the stromboids is the spiral twist of the shell in the part farthest away from the head; they are also furnished from birth with an operculum. And, further, all stromboid testaceans have their shells on the right hand side, and move not in the direction of the spire, but the opposite way. Such are the diversities observed in the external parts of these animals.

The internal structure is almost the same in all these creatures, and in the stromboids especially; for it is in size that these latter differ from one another, and in accidents of the nature of excess or defect. And there is not much difference between most of the univalves and bivalves; but, while those that open and shut differ from one another but slightly, they differ considerably from such as are incapable of motion. And this will be illustrated more satisfactorily hereafter.

The spiral-shaped testaceans are all similarly constructed, but differ from one another, as has been said, in the way of excess or defect (for the larger species have larger and more conspicuous organs, and the smaller have smaller and less conspicuous), and, furthermore, in relative hardness or softness, and in other such accidents or properties. All the stromboids, for instance, have the flesh that extrudes from the mouth of the shell, hard and stiff; some more, and some less. From the middle of this protrudes the head and two horns, and these horns are large in the large species, but exceedingly minute in the smaller ones. The head protrudes from them all in the same way; and, if the animal be alarmed, the head draws in again. Some of these creatures have a mouth and teeth, as the snail; teeth sharp, and small, and delicate. They have also a proboscis just like that of the fly; and the proboscis is tongue-shaped. The ceryx and the purple murex have this organ firm and solid; and just as the myops, or horse-fly, and the oestrus, or gadfly, can pierce the skin of a quadruped, so is that proboscis proportionately stronger in these testaceans; for they bore right through the shells of other shell-fish on which they prey. The stomach follows close upon the mouth, and, by the way, this organ in the snail resembles a bird's crop. Underneath come two white firm formations, mastoid or papillary in form; and similar formations are found in the cuttle-fish also, only that they are of a firmer consistency in the cuttle-fish. After the stomach comes an esophagus, simple and long, extending to the poppy or quasi-liver, which is in the innermost recess of the shell. All these statements may be verified in the case of the purple murex and the ceryx by observation within the whorl of the shell. What comes next to the esophagus is the gut; in fact, the gut is continuous with the esophagus, and runs its whole length uncomplicated to the outlet of the residuum. The gut has its point of origin in the region of the coil of the mecon, or so-called 'poppy', and is wider hereabouts (for remember, the mecon is for the most part a sort of excretion in all testaceans); it then takes a bend and runs up again towards the fleshy part, and terminates by the side of the head, where the animal discharges its residuum; and this holds good in the case of all stromboid testaceans, whether terrestrial or marine. From the stomach there is drawn in a parallel direction with the esophagus, in the larger snails, a long white duct enveloped in a membrane, resembling in colour the mastoid formations higher up; and in it are nicks or interruptions, as in the egg-mass of the crawfish, only, by the way, the duct of which we are treating is white and the egg-mass of the crawfish is red. This formation has no outlet nor duct, but is enveloped in a thin membrane with a narrow cavity in its interior. And from the gut downward extend black and rough formations, in close connexion, something like the formations in the tortoise, only not so black. Marine snails, also, have these formations, and the white ones, only that the formations are smaller in the smaller species.

The non-spiral univalves and bivalves are in some respect similar in construction, and in some respects dissimilar, to the spiral testaceans. They all have a head and horns, and a mouth, and the organ resembling a tongue; but these organs, in the smaller species, are indiscernible owing to the minuteness of these animals, and some are indiscernible even in the larger species when dead, or when at rest and motionless. They all have the mecon, or poppy, but not all in the same place, nor of equal size, nor similarly open to observation; thus, the limpets have this organ deep down in the bottom of the shell, and the bivalves at the hinge connecting the two valves. They also have in all cases the hairy growths or beards, in a circular form, as in the scallops. And, with regard to the so-called 'egg', in those that have it, when they have it, it is situated in one of the semi-circles of the periphery, as is the case with the white formation in the snail; for this white formation in the snail corresponds to the so-called egg of which we are speaking. But all these organs, as has been stated, are distinctly traceable in the larger species, while in the small ones they are in some cases almost, and in others altogether, indiscernible. Hence they are most plainly visible in the large scallops; and these are the bivalves that have one valve flat-shaped, like the lid of a pot. The outlet of the excretion is in all these animals (save for the exception to be afterwards related) on one side; for there is a passage whereby the excretion passes out. (And, remember, the mecon or poppy, as has been stated, is an excretion in all these animals-an excretion enveloped in a membrane.) The so-called egg has no outlet in any of these creatures, but is merely an excrescence in the fleshy mass; and it is not situated in the same region with the gut, but the 'egg' is situated on the right-hand side and the gut on the left. Such are the relations of the anal vent in most of these animals; but in the case of the wild limpet (called by some the 'sea-ear'), the residuum issues beneath the shell, for the shell is perforated to give an outlet. In this particular limpet the stomach is seen coming after the mouth, and the egg-shaped formations are discernible. But for the relative positions of these parts you are referred to my Treatise on Anatomy.

The so-called carcinium or hermit crab is in a way intermediate between the crustaceans and the testaceans. In its nature it resembles the crawfish kind, and it is born simple of itself, but by its habit of introducing itself into a shell and living there it resembles the testaceans, and so appears to partake of the characters of both kinds. In shape, to give a simple illustration, it resembles a spider, only that the part below the head and thorax is larger in this creature than in the spider. It has two thin red horns, and underneath these horns two long eyes, not retreating inwards, nor turning sideways like the eyes of the crab, but protruding straight out; and underneath these eyes the mouth, and round about the mouth several hair-like growths, and next after these two bifurcate legs or claws, whereby it draws in objects towards itself, and two other legs on either side, and a third small one. All below the thorax is soft, and when opened in dissection is found to be sallow-coloured within. From the mouth there runs a single passage right on to the stomach, but the passage for the excretions is not discernible. The legs and the thorax are hard, but not so hard as the legs and the thorax of the crab. It does not adhere to its shell like the purple murex and the ceryx, but can easily slip out of it. It is longer when found in the shell of the stromboids than when found in the shell of the neritae.

And, by the way, the animal found in the shell of the neritae is a separate species, like to the other in most respects; but of its bifurcate feet or claws, the right-hand one is small and the left-hand one is large, and it progresses chiefly by the aid of this latter and larger one. (In the shells of these animals, and in certain others, there is found a parasite whose mode of attachment is similar. The particular one which we have just described is named the cyllarus.)

The nerites has a smooth large round shell, and resembles the ceryx in shape, only the poppy-juice is, in its case, not black but red. It clings with great force near the middle. In calm weather, then, they go free afield, but when the wind blows the carcinia take shelter against the rocks: the neritae themselves cling fast like limpets; and the same is the case with the hemorrhoid or aporrhaid and all others of the like kind. And, by the way, they cling to the rock, when they turn back their operculum, for this operculum seems like a lid; in fact this structure represents the one part, in the stromboids, of that which in the bivalves is a duplicate shell. The interior of the animal is fleshy, and the mouth is inside. And it is the same with the hemorrhoid, the purple murex, and all suchlike animals.

Such of the little crabs as have the left foot or claw the bigger of the two are found in the neritae, but not in the stromboids. are some snail-shells which have inside them creatures resembling those little crayfish that are also found in fresh water. These creatures, however, differ in having the part inside the shells But as to the characters, you are referred to my Treatise on Anatomy.

Part 5

The urchins are devoid of flesh, and this is a character peculiar to them; and while they are in all cases empty and devoid of any flesh within, they are in all cases furnished with the black formations. There are several species of the urchin, and one of these is that which is made use of for food; this is the kind in which are found the so-called eggs, large and edible, in the larger and smaller specimens alike; for even when as yet very small they are provided with them. There are two other species, the spatangus, and the so-called bryssus, these animals are pelagic and scarce. Further, there are the echinometrae, or 'mother-urchins', the largest in size of all the species. In addition to these there is another species, small in size, but furnished with large hard spines; it lives in the sea at a depth of several fathoms; and is used by some people as a specific for cases of strangury. In the neighbourhood of Torone there are sea-urchins of a white colour, shells, spines, eggs and all, and that are longer than the ordinary sea-urchin. The spine in this species is not large nor strong, but rather limp; and the black formations in connexion with the mouth are more than usually numerous, and communicate with the external duct, but not with one another; in point of fact, the animal is in a manner divided up by them. The edible urchin moves with greatest freedom and most often; and this is indicated by the fact that these urchins have always something or other on their spines.

All urchins are supplied with eggs, but in some of the species the eggs are exceedingly small and unfit for food. Singularly enough, the urchin has what we may call its head and mouth down below, and a place for the issue of the residuum up above; (and this same property is common to all stromboids and to limpets). For the food on which the creature lives lies down below; consequently the mouth has a position well adapted for getting at the food, and the excretion is above, near to the back of the shell. The urchin has, also, five hollow teeth inside, and in the middle of these teeth a fleshy substance serving the office of a tongue. Next to this comes the esophagus, and then the stomach, divided into five parts, and filled with excretion, all the five parts uniting at the anal vent, where the shell is perforated for an outlet. Underneath the stomach, in another membrane, are the so-called eggs, identical in number in all cases, and that number is always an odd number, to wit five. Up above, the black formations are attached to the starting-point of the teeth, and they are bitter to the taste, and unfit for food. A similar or at least an analogous formation is found in many animals; as, for instance, in the tortoise, the toad, the frog, the stromboids, and, generally, in the molluscs; but the formation varies here and there in colour, and in all cases is altogether uneatable, or more or less unpalatable. In reality the mouth-apparatus of the urchin is continuous from one end to the other, but to outward appearance it is not so, but looks like a horn lantern with the panes of horn left out. The urchin uses its spines as feet; for it rests its weight on these, and then moving shifts from place to place.

Part 6

The so-called tethyum or ascidian has of all these animals the most remarkable characteristics. It is the only mollusc that has its entire body concealed within its shell, and the shell is a substance intermediate between hide and shell, so that it cuts like a piece of hard leather. It is attached to rocks by its shell, and is provided with two passages placed at a distance from one another, very minute and hard to see, whereby it admits and discharges the sea-water; for it has no visible excretion (whereas of shell fish in general some resemble the urchin in this matter of excretion, and others are provided with the so-called mecon, or poppy-juice). If the animal be opened, it is found to have, in the first place, a tendinous membrane running round inside the shell-like substance, and within this membrane is the flesh-like substance of the ascidian, not resembling that in other molluscs; but this flesh, to which I now allude, is the same in all ascidia. And this substance is attached in two places to the membrane and the skin, obliquely; and at the point of attachment the space is narrowed from side to side, where the fleshy substance stretches towards the passages that lead outwards through the shell; and here it discharges and admits food and liquid matter, just as it would if one of the passages were a mouth and the other an anal vent; and one of the passages is somewhat wider than the other Inside it has a pair of cavities, one on either side, a small partition separating them; and one of these two cavities contains the liquid. The creature has no other organ whether motor or sensory, nor, as was said in the case of the others, is it furnished with any organ connected with excretion, as other shell-fish are. The colour of the ascidian is in some cases sallow, and in other cases red.

There is, furthermore, the genus of the sea-nettles, peculiar in its way. The sea-nettle, or sea-anemone, clings to rocks like certain of the testaceans, but at times relaxes its hold. It has no shell, but its entire body is fleshy. It is sensitive to touch, and, if you put your hand to it, it will seize and cling to it, as the cuttlefish would do with its feelers, and in such a way as to make the flesh of your hand swell up. Its mouth is in the centre of its body, and it lives adhering to the rock as an oyster to its shell. If any little fish come up against it, it clings to it; in fact, just as I described it above as doing to your hand, so it does to anything edible that comes in its way; and it feeds upon sea-urchins and scallops. Another species of the sea-nettle roams freely abroad. The sea-nettle appears to be devoid altogether of excretion, and in this respect it resembles a plant.

Of sea-nettles there are two species, the lesser and more edible, and the large hard ones, such as are found in the neighbourhood of Chalcis. In winter time their flesh is firm, and accordingly they are sought after as articles of food, but in summer weather they are worthless, for they become thin and watery, and if you catch at them they break at once into bits, and cannot be taken off the rocks entire; and being oppressed by the heat they tend to slip back into the crevices of the rocks.

So much for the external and the internal organs of molluscs, crustaceans, and testaceans.

Part 7

We now proceed to treat of insects in like manner. This genus comprises many species, and, though several kinds are clearly related to one another, these are not classified under one common designation, as in the case of the bee, the drone, the wasp, and all such insects, and again as in the case of those that have their wings in a sheath or shard, like the cockchafer, the carabus or stag-beetle, the cantharis or blister-beetle, and the like.

Insects have three parts common to them all; the head, the trunk containing the stomach, and a third part in betwixt these two, corresponding to what in other creatures embraces chest and back. In the majority of insects this intermediate part is single; but in the long and multipedal insects it has practically the same number of segments as of nicks.

All insects when cut in two continue to live, excepting such as are naturally cold by nature, or such as from their minute size chill rapidly; though, by the way, wasps notwithstanding their small size continue living after severance. In conjunction with the middle portion either the head or the stomach can live, but the head cannot live by itself. Insects that are long in shape and many-footed can live for a long while after being cut in twain, and the severed portions can move in either direction, backwards or forwards; thus, the hinder portion, if cut off, can crawl either in the direction of the section or in the direction of the tail, as is observed in the scolopendra.

All insects have eyes, but no other organ of sense discernible, except that some insects have a kind of a tongue corresponding to a similar organ common to all testaceans; and by this organ such insects taste and imbibe their food. In some insects this organ is soft; in other insects it is firm; as it is, by the way, in the purple-fish, among testaceans. In the horsefly and the gadfly this organ is hard, and indeed it is hard in most insects. In point of fact, such insects as have no sting in the rear use this organ as a weapon, (and, by the way, such insects as are provided with this organ are unprovided with teeth, with the exception of a few insects); the fly by a touch can draw blood with this organ, and the gnat can prick or sting with it.

Certain insects are furnished with prickers or stings. Some insects have the sting inside, as the bee and the wasp, others outside, as the scorpion; and, by the way, this is the only insect furnished with a long tail. And, further, the scorpion is furnished with claws, as is also the creature resembling a scorpion found within the pages of books.

In addition to their other organs, flying insects are furnished with wings. Some insects are dipterous or double-winged, as the fly; others are tetrapterous or furnished with four wings, as the bee; and, by the way, no insect with only two wings has a sting in the rear. Again, some winged insects have a sheath or shard for their wings, as the cockchafer; whereas in others the wings are unsheathed, as in the bee. But in the case of all alike, flight is in no way modified by tail-steerage, and the wing is devoid of quill-structure or division of any kind.

Again, some insects have antennae in front of their eyes, as the butterfly and the horned beetle. Such of them as have the power of jumping have the hinder legs the longer; and these long hind-legs whereby they jump bend backwards like the hind-legs of quadrupeds. All insects have the belly different from the back; as, in fact, is the case with all animals. The flesh of an insect's body is neither shell-like nor is it like the internal substance of shell-covered animals, nor is it like flesh in the ordinary sense of the term; but it is a something intermediate in quality. Wherefore they have nor spine, nor bone, nor sepia-bone, nor enveloping shell; but their body by its hardness is its own protection and requires no extraneous support. However, insects have a skin; but the skin is exceedingly thin. These and such-like are the external organs of insects.

Internally, next after the mouth, comes a gut, in the majority of cases straight and simple down to the outlet of the residuum: but in a few cases the gut is coiled. No insect is provided with any viscera, or is supplied with fat; and these statements apply to all animals devoid of blood. Some have a stomach also, and attached to this the rest of the gut, either simple or convoluted as in the case of the acris or grasshopper.

The tettix or cicada, alone of such creatures (and, in fact, alone of all creatures), is unprovided with a mouth, but it is provided with the tongue-like formation found in insects furnished with frontward stings; and this formation in the cicada is long, continuous, and devoid of any split; and by the aid of this the creature feeds on dew, and on dew only, and in its stomach no excretion is ever found. Of the cicada there are several kinds, and they differ from one another in relative magnitude, and in this respect that the achetes or chirper is provided with a cleft or aperture under the hypozoma and has in it a membrane quite discernible, whilst the membrane is indiscernible in the tettigonia.

Furthermore, there are some strange creatures to be found in the sea, which from their rarity we are unable to classify. Experienced fishermen affirm, some that they have at times seen in the sea animals like sticks, black, rounded, and of the same thickness throughout; others that they have seen creatures resembling shields, red in colour, and furnished with fins packed close together; and others that they have seen creatures resembling the male organ in shape and size, with a pair of fins in the place of the testicles, and they aver that on one occasion a creature of this description was brought up on the end of a nightline.

So much then for the parts, external and internal, exceptional and common, of all animals.

Part 8

We now proceed to treat of the senses; for there are diversities in animals with regard to the senses, seeing that some animals have the use of all the senses, and others the use of a limited number of them. The total number of the senses (for we have no experience of any special sense not here included), is five: sight, hearing, smell, taste, and touch.

Man, then, and all vivipara that have feet, and, further, all red-blooded ovipara, appear to have the use of all the five senses, except where some isolated species has been subjected to mutilation, as in the case of the mole. For this animal is deprived of sight; it has no eyes visible, but if the skin-a thick one, by the way-be stripped off the head, about the place in the exterior where eyes usually are, the eyes are found inside in a stunted condition, furnished with all the parts found in ordinary eyes; that is to say, we find there the black rim, and the fatty part surrounding it; but all these parts are smaller than the same parts in ordinary visible eyes. There is no external sign of the existence of these organs in the mole, owing to the thickness of the skin drawn over them, so that it would seem that the natural course of development were congenitally arrested; (for extending from the brain at its junction with the marrow are two strong sinewy ducts running past the sockets of the eyes, and terminating at the upper eye-teeth). All the other animals of the kinds above mentioned have a perception of colour and of sound, and the senses of smell and taste; the fifth sense, that, namely, of touch, is common to all animals whatsoever.

In some animals the organs of sense are plainly discernible; and this is especially the case with the eyes. For animals have a special locality for the eyes, and also a special locality for hearing: that is to say, some animals have ears, while others have the passage for sound discernible. It is the same with the sense of smell; that is to say, some animals have nostrils, and others have only the passages for smell, such as birds. It is the same also with the organ of taste, the tongue. Of aquatic red-blooded animals, fishes possess the organ of taste, namely the tongue, but it is in an imperfect and amorphous form, in other words it is osseous and undetached. In some fish the palate is fleshy, as in the fresh-water carp, so that by an inattentive observer it might be mistaken for a tongue.

There is no doubt but that fishes have the sense of taste, for a great number of them delight in special flavours; and fishes freely take the hook if it be baited with a piece of flesh from a tunny or from any fat fish, obviously enjoying the taste and the eating of food of this kind. Fishes have no visible organs for hearing or for smell; for what might appear to indicate an organ for smell in the region of the nostril has no communication with the brain. These indications, in fact, in some cases lead nowhere, like blind alleys, and in other cases lead only to the gills; but for all this fishes undoubtedly hear and smell. For they are observed to run away from any loud noise, such as would be made by the rowing of a galley, so as to become easy of capture in their holes; for, by the way, though a sound be very slight in the open air, it has a loud and alarming resonance to creatures that hear under water. And this is shown in the capture of the dolphin; for when the hunters have enclosed a shoal of these fishes with a ring of their canoes, they set up from inside the canoes a loud splashing in the water, and by so doing induce the creatures to run in a shoal high and dry up on the beach, and so capture them while stupefied with the noise. And yet, for all this, the dolphin has no organ of hearing discernible. Furthermore, when engaged in their craft, fishermen are particularly careful to make no noise with oar or net; and after they have spied a shoal, they let down their nets at a spot so far off that they count upon no noise being likely to reach the shoal, occasioned either by oar or by the surging of their boats through the water; and the crews are strictly enjoined to preserve silence until the shoal has been surrounded. And, at times, when they want the fish to crowd together, they adopt the stratagem of the dolphin-hunter; in other words they clatter stones together, that the fish may, in their fright, gather close into one spot, and so they envelop them within their nets. (Before surrounding them, then, they preserve silence, as was said; but, after hemming the shoal in, they call on every man to shout out aloud and make any kind of noise; for on hearing the noise and hubbub the fish are sure to tumble into the nets from sheer fright.) Further, when fishermen see a shoal of fish feeding at a distance, disporting themselves in calm bright weather on the surface of the water, if they are anxious to descry the size of the fish and to learn what kind of a fish it is, they may succeed in coming upon the shoal whilst yet basking at the surface if they sail up without the slightest noise, but if any man make a noise previously, the shoal will be seen to scurry away in alarm. Again, there is a small river-fish called the cottus or bullhead; this creature burrows under a rock, and fishers catch it by clattering stones against the rock, and the fish, bewildered at the noise, darts out of its hiding-place. From these facts it is quite obvious that fishes can hear; and indeed some people, from living near the sea and frequently witnessing such phenomena, affirm that of all living creatures the fish is the quickest of hearing. And, by the way, of all fishes the quickest of hearing are the cestreus or mullet, the chremps, the labrax or bass, the salpe or saupe, the chromis or sciaena, and such like. Other fishes are less quick of hearing, and, as might be expected, are more apt to be found living at the bottom of the sea.

The case is similar in regard to the sense of smell. Thus, as a rule, fishes will not touch a bait that is not fresh, neither are they all caught by one and the same bait, but they are severally caught by baits suited to their several likings, and these baits they distinguish by their sense of smell; and, by the way, some fishes are attracted by malodorous baits, as the saupe, for instance, is attracted by excrement. Again, a number of fishes live in caves; and accordingly fishermen, when they want to entice them out, smear the mouth of a cave with strong-smelling pickles, and the fish are Soon attracted to the smell. And the eel is caught in a similar way; for the fisherman lays down an earthen pot that has held pickles, after inserting a 'weel' in the neck thereof. As a general rule, fishes are especially attracted by savoury smells. For this reason, fishermen roast the fleshy parts of the cuttle-fish and use it as bait on account of its smell, for fish are peculiarly attracted by it; they also bake the octopus and bait their fish-baskets or weels with it, entirely, as they say, on account of its smell. Furthermore, gregarious fishes, if fish washings or bilge-water be thrown overboard, are observed to scud off to a distance, from apparent dislike of the smell. And it is asserted that they can at once detect by smell the presence of their own blood; and this faculty is manifested by their hurrying off to a great distance whenever fish-blood is spilt in the sea. And, as a general rule, if you bait your weel with a stinking bait, the fish refuse to enter the weel or even to draw near; but if you bait the weel with a fresh and savoury bait, they come at once from long distances and swim into it. And all this is particularly manifest in the dolphin; for, as was stated, it has no visible organ of hearing, and yet it is captured when stupefied with noise; and so, while it has no visible organ for smell, it has the sense of smell remarkably keen. It is manifest, then, that the animals above mentioned are in possession of all the five senses.

All other animals may, with very few exceptions, be comprehended within four genera: to wit, molluscs, crustaceans, testaceans, and insects. Of these four genera, the mollusc, the crustacean, and the insect have all the senses: at all events, they have sight, smell, and taste. As for insects, both winged and wingless, they can detect the presence of scented objects afar off, as for instance bees and snipes detect the presence of honey at a distance; and do so recognizing it by smell. Many insects are killed by the smell of brimstone; ants, if the apertures to their dwellings be smeared with powdered origanum and brimstone, quit their nests; and most insects may be banished with burnt hart's horn, or better still by the burning of the gum styrax. The cuttle-fish, the octopus, and the crawfish may be caught by bait. The octopus, in fact, clings so tightly to the rocks that it cannot be pulled off, but remains attached even when the knife is employed to sever it; and yet, if you apply fleabane to the creature, it drops off at the very smell of it. The facts are similar in regard to taste. For the food that insects go in quest of is of diverse kinds, and they do not all delight in the same flavours: for instance, the bee never settles on a withered or wilted flower, but on fresh and sweet ones; and the conops or gnat settles only on acrid substances and not on sweet. The sense of touch, by the way, as has been remarked, is common to all animals. Testaceans have the senses of smell and taste. With regard to their possession of the sense of smell, that is proved by the use of baits, e.g. in the case of the purple-fish; for this creature is enticed by baits of rancid meat, which it perceives and is attracted to from a great distance. The proof that it possesses a sense of taste hangs by the proof of its sense of smell; for whenever an animal is attracted to a thing by perceiving its smell, it is sure to like the taste of it. Further, all animals furnished with a mouth derive pleasure or pain from the touch of sapid juices.

With regard to sight and hearing, we cannot make statements with thorough confidence or on irrefutable evidence. However, the solen or razor-fish, if you make a noise, appears to burrow in the sand, and to hide himself deeper when he hears the approach of the iron rod (for the animal, be it observed, juts a little out of its hole, while the greater part of the body remains within),-and scallops, if you present your finger near their open valves, close them tight again as though they could see what you were doing. Furthermore, when fishermen are laying bait for neritae, they always get to leeward of them, and never speak a word while so engaged, under the firm impression that the animal can smell and hear; and they assure us that, if any one speaks aloud, the creature makes efforts to escape. With regard to testaceans, of the walking or creeping species the urchin appears to have the least developed sense of smell; and, of the stationary species, the ascidian and the barnacle.

So much for the organs of sense in the general run of animals. We now proceed to treat of voice.

Part 9

Voice and sound are different from one another; and language differs from voice and sound. The fact is that no animal can give utterance to voice except by the action of the pharynx, and consequently such animals as are devoid of lung have no voice; and language is the articulation of vocal sounds by the instrumentality of the tongue. Thus, the voice and larynx can emit vocal or vowel sounds; non-vocal or consonantal sounds are made by the tongue and the lips; and out of these vocal and non-vocal sounds language is composed. Consequently, animals that have no tongue at all or that have a tongue not freely detached, have neither voice nor language; although, by the way, they may be enabled to make noises or sounds by other organs than the tongue.

Insects, for instance, have no voice and no language, but they can emit sound by internal air or wind, though not by the emission of air or wind; for no insects are capable of respiration. But some of them make a humming noise, like the bee and the other winged insects; and others are said to sing, as the cicada. And all these latter insects make their special noises by means of the membrane that is underneath the 'hypozoma'-those insects, that is to say, whose body is thus divided; as for instance, one species of cicada, which makes the sound by means of the friction of the air. Flies and bees, and the like, produce their special noise by opening and shutting their wings in the act of flying; for the noise made is by the friction of air between the wings when in motion. The noise made by grasshoppers is produced by rubbing or reverberating with their long hind-legs.

No mollusc or crustacean can produce any natural voice or sound. Fishes can produce no voice, for they have no lungs, nor windpipe and pharynx; but they emit certain inarticulate sounds and squeaks, which is what is called their 'voice', as the lyra or gurnard, and the sciaena (for these fishes make a grunting kind of noise) and the caprus or boar-fish in the river Achelous, and the chalcis and the cuckoo-fish; for the chalcis makes a sort piping sound, and the cuckoo-fish makes a sound greatly like the cry of the cuckoo, and is nicknamed from the circumstance. The apparent voice in all these fishes is a sound caused in some cases by a rubbing motion of their gills, which by the way are prickly, or in other cases by internal parts about their bellies; for they all have air or wind inside them, by rubbing and moving which they produce the sounds. Some cartilaginous fish seem to squeak.

But in these cases the term 'voice' is inappropriate; the more correct expression would be 'sound'. For the scallop, when it goes along supporting itself on the water, which is technically called 'flying', makes a whizzing sound; and so does the sea-swallow or flying-fish: for this fish flies in the air, clean out of the water, being furnished with fins broad and long. Just then as in the flight of birds the sound made by their wings is obviously not voice, so is it in the case of all these other creatures.

The dolphin, when taken out of the water, gives a squeak and moans in the air, but these noises do not resemble those above mentioned. For this creature has a voice (and can therefore utter vocal or vowel sounds), for it is furnished with a lung and a windpipe; but its tongue is not loose, nor has it lips, so as to give utterance to an articulate sound (or a sound of vowel and consonant in combination.)

Of animals which are furnished with tongue and lung, the oviparous quadrupeds produce a voice, but a feeble one; in some cases, a shrill piping sound, like the serpent; in others, a thin faint cry; in others, a low hiss, like the tortoise. The formation of the tongue in the frog is exceptional. The front part of the tongue, which in other animals is detached, is tightly fixed in the frog as it is in all fishes; but the part towards the pharynx is freely detached, and may, so to speak, be spat outwards, and it is with this that it makes its peculiar croak. The croaking that goes on in the marsh is the call of the males to the females at rutting time; and, by the way, all animals have a special cry for the like end at the like season, as is observed in the case of goats, swine, and sheep. (The bull-frog makes its croaking noise by putting its under jaw on a level with the surface of the water and extending its upper jaw to its utmost capacity. The tension is so great that the upper jaw becomes transparent, and the animal's eyes shine through the jaw like lamps; for, by the way, the commerce of the sexes takes place usually in the night time.) Birds can utter vocal sounds; and such of them can articulate best as have the tongue moderately flat, and also such as have thin delicate tongues. In some cases, the male and the female utter the same note; in other cases, different notes. The smaller birds are more vocal and given to chirping than the larger ones; but in the pairing season every species of bird becomes particularly vocal. Some of them call when fighting, as the quail, others cry or crow when challenging to combat, as the partridge, or when victorious, as the barn-door cock. In some cases cock-birds and hens sing alike, as is observed in the nightingale, only that the hen stops singing when brooding or rearing her young; in other birds, the cocks sing more than the hens; in fact, with barn-door fowls and quails, the cock sings and the hen does not.

Viviparous quadrupeds utter vocal sounds of different kinds, but they have no power of converse. In fact, this power, or language, is peculiar to man. For while the capability of talking implies the capability of uttering vocal sounds, the converse does not hold good. Men that are born deaf are in all cases also dumb; that is, they can make vocal sounds, but they cannot speak. Children, just as they have no control over other parts, so have no control, at first, over the tongue; but it is so far imperfect, and only frees and detaches itself by degrees, so that in the interval children for the most part lisp and stutter.

Vocal sounds and modes of language differ according to locality. Vocal sounds are characterized chiefly by their pitch, whether high or low, and the kinds of sound capable of being produced are identical within the limits of one and the same species; but articulate sound, that one might reasonably designate 'language', differs both in various animals, and also in the same species according to diversity of locality; as for instance, some partridges cackle, and some make a shrill twittering noise. Of little birds, some sing a different note from the parent birds, if they have been removed from the nest and have heard other birds singing; and a mother-nightingale has been observed to give lessons in singing to a young bird, from which spectacle we might obviously infer that the song of the bird was not equally congenital with mere voice, but was something capable of modification and of improvement. Men have the same voice or vocal sounds, but they differ from one another in speech or language.

The elephant makes a vocal sound of a windlike sort by the mouth alone, unaided by the trunk, just like the sound of a man panting or sighing; but, if it employ the trunk as well, the sound produced is like that of a hoarse trumpet.

Part 10

With regard to the sleeping and waking of animals, all creatures that are red-blooded and provided with legs give sensible proof that they go to sleep and that they waken up from sleep; for, as a matter of fact, all animals that are furnished with eyelids shut them up when they go to sleep. Furthermore, it would appear that not only do men dream, but horses also, and dogs, and oxen; aye, and sheep, and goats, and all viviparous quadrupeds; and dogs show their dreaming by barking in their sleep. With regard to oviparous animals we cannot be sure that they dream, but most undoubtedly they sleep. And the same may be said of water animals, such as fishes, molluscs, crustaceans, to wit crawfish and the like. These animals sleep without doubt, although their sleep is of very short duration. The proof of their sleeping cannot be got from the condition of their eyes-for none of these creatures are furnished with eyelids-but can be obtained only from their motionless repose.

Apart from the irritation caused by lice and what are nicknamed fleas, fish are met with in a state so motionless that one might easily catch them by hand; and, as a matter of fact, these little creatures, if the fish remain long in one position, will attack them in myriads and devour them. For these parasites are found in the depths of the sea, and are so numerous that they devour any bait made of fish's flesh if it be left long on the ground at the bottom; and fishermen often draw up a cluster of them, all clinging on to the bait.

But it is from the following facts that we may more reasonably infer that fishes sleep. Very often it is possible to take a fish off its guard so far as to catch hold of it or to give it a blow unawares; and all the while that you are preparing to catch or strike it, the fish is quite still but for a slight motion of the tail. And it is quite obvious that the animal is sleeping, from its movements if any disturbance be made during its repose; for it moves just as you would expect in a creature suddenly awakened. Further, owing to their being asleep, fish may be captured by torchlight. The watchmen in the tunny-fishery often take advantage of the fish being asleep to envelop them in a circle of nets; and it is quite obvious that they were thus sleeping by their lying still and allowing the glistening under-parts of their bodies to become visible, while the capture is taking Place. They sleep in the night-time more than during the day; and so soundly at night that you may cast the net without making them stir. Fish, as a general rule, sleep close to the ground, or to the sand or to a stone at the bottom, or after concealing themselves under a rock or the ground. Flat fish go to sleep in the sand; and they can be distinguished by the outlines of their shapes in the sand, and are caught in this position by being speared with pronged instruments. The bass, the chrysophrys or gilt-head, the mullet, and fish of the like sort are often caught in the daytime by the prong owing to their having been surprised when sleeping; for it is scarcely probable that fish could be pronged while awake. Cartilaginous fish sleep at times so soundly that they may be caught by hand. The dolphin and the whale, and all such as are furnished with a blow-hole, sleep with the blow-hole over the surface of the water, and breathe through the blow-hole while they keep up a quiet flapping of their fins; indeed, some mariners assure us that they have actually heard the dolphin snoring.

Molluscs sleep like fish, and crustaceans also. It is plain also that insects sleep; for there can be no mistaking their condition of motionless repose. In the bee the fact of its being asleep is very obvious; for at night-time bees are at rest and cease to hum. But the fact that insects sleep may be very well seen in the case of common every-day creatures; for not only do they rest at night-time from dimness of vision (and, by the way, all hard-eyed creatures see but indistinctly), but even if a lighted candle be presented they continue sleeping quite as soundly.

Of all animals man is most given to dreaming. Children and infants do not dream, but in most cases dreaming comes on at the age of four or five years. Instances have been known of full-grown men and women that have never dreamed at all; in exceptional cases of this kind, it has been observed that when a dream occurs in advanced life it prognosticates either actual dissolution or a general break-up of the system.

So much then for sensation and for the phenomena of sleeping and of awakening.

Part 11

With regard to sex, some animals are divided into male and female, but others are not so divided but can only be said in a comparative way to bring forth young and to be pregnant. In animals that live confined to one spot there is no duality of sex; nor is there such, in fact, in any testaceans. In molluscs and in crustaceans we find male and female: and, indeed, in all animals furnished with feet, biped or quadruped; in short, in all such as by copulation engender either live young or egg or grub. In the several genera, with however certain exceptions, there either absolutely is or absolutely is not a duality of sex. Thus, in quadrupeds the duality is universal, while the absence of such duality is universal in testaceans, and of these creatures, as with plants, some individuals are fruitful and some are not their lying still

But among insects and fishes, some cases are found wholly devoid of this duality of sex. For instance, the eel is neither male nor female, and can engender nothing. In fact, those who assert that eels are at times found with hair-like or worm-like progeny attached, make only random assertions from not having carefully noticed the locality of such attachments. For no eel nor animal of this kind is ever viviparous unless previously oviparous; and no eel was ever yet seen with an egg. And animals that are viviparous have their young in the womb and closely attached, and not in the belly; for, if the embryo were kept in the belly, it would be subjected to the process of digestion like ordinary food. When people rest duality of sex in the eel on the assertion that the head of the male is bigger and longer, and the head of the female smaller and more snubbed, they are taking diversity of species for diversity of sex.

There are certain fish that are nicknamed the epitragiae, or capon-fish, and, by the way, fish of this description are found in fresh water, as the carp and the balagrus. This sort of fish never has either roe or milt; but they are hard and fat all over, and are furnished with a small gut; and these fish are regarded as of super-excellent quality.

Again, just as in testaceans and in plants there is what bears and engenders, but not what impregnates, so is it, among fishes, with the psetta, the erythrinus, and the channe; for these fish are in all cases found furnished with eggs.

As a general rule, in red-blooded animals furnished with feet and not oviparous, the male is larger and longer-lived than the female (except with the mule, where the female is longer-lived and bigger than the male); whereas in oviparous and vermiparous creatures, as in fishes and in insects, the female is larger than the male; as, for instance, with the serpent, the phalangium or venom-spider, the gecko, and the frog. The same difference in size of the sexes is found in fishes, as, for instance, in the smaller cartilaginous fishes, in the greater part of the gregarious species, and in all that live in and about rocks. The fact that the female is longer-lived than the male is inferred from the fact that female fishes are caught older than males. Furthermore, in all animals the upper and front parts are better, stronger, and more thoroughly equipped in the male than in the female, whereas in the female those parts are the better that may be termed hinder-parts or underparts. And this statement is applicable to man and to all vivipara that have feet. Again, the female is less muscular and less compactly jointed, and more thin and delicate in the hair-that is, where hair is found; and, where there is no hair, less strongly furnished in some analogous substance. And the female is more flaccid in texture of flesh, and more knock-kneed, and the shin-bones are thinner; and the feet are more arched and hollow in such animals as are furnished with feet. And with regard to voice, the female in all animals that are vocal has a thinner and sharper voice than the male; except, by the way, with kine, for the lowing and bellowing of the cow has a deeper note than that of the bull. With regard to organs of defence and offence, such as teeth, tusks, horns, spurs, and the like, these in some species the male possesses and the female does not; as, for instance, the hind has no horns, and where the cock-bird has a spur the hen is entirely destitute of the organ; and in like manner the sow is devoid of tusks. In other species such organs are found in both sexes, but are more perfectly developed in the male; as, for instance, the horn of the bull is more powerful than the horn of the cow.

Book V

Part 1

AS to the parts internal and external that all animals are furnished withal, and further as to the senses, to voice, and sleep, and the duality sex, all these topics have now been touched upon. It now remains for us to discuss, duly and in order, their several modes of propagation.

These modes are many and diverse, and in some respects are like, and in other respects are unlike to one another. As we carried on our previous discussion genus by genus, so we must attempt to follow the same divisions in our present argument; only that whereas in the former case we started with a consideration of the parts of man, in the present case it behooves us to treat of man last of all because he involves most discussion. We shall commence, then, with testaceans, and then proceed to crustaceans, and then to the other genera in due order; and these other genera are, severally, molluscs, and insects, then fishes viviparous and fishes oviparous, and next birds; and afterwards we shall treat of animals provided with feet, both such as are oviparous and such as are viviparous, and we may observe that some quadrupeds are viviparous, but that the only viviparous biped is man.

Now there is one property that animals are found to have in common with plants. For some plants are generated from the seed of plants, whilst other plants are self-generated through the formation of some elemental principle similar to a seed; and of these latter plants some derive their nutriment from the ground, whilst others grow inside other plants, as is mentioned, by the way, in my treatise on Botany. So with animals, some spring from parent animals according to their kind, whilst others grow spontaneously and not from kindred stock; and of these instances of spontaneous generation some come from putrefying earth or vegetable matter, as is the case with a number of insects, while others are spontaneously generated in the inside of animals out of the secretions of their several organs.

In animals where generation goes by heredity, wherever there is duality of sex generation is due to copulation. In the group of fishes, however, there are some that are neither male nor female, and these, while they are identical generically with other fish, differ from them specifically; but there are others that stand altogether isolated and apart by themselves. Other fishes there are that are always female and never male, and from them are conceived what correspond to the wind-eggs in birds. Such eggs, by the way, in birds are all unfruitful; but it is their nature to be independently capable of generation up to the egg-stage, unless indeed there be some other mode than the one familiar to us of intercourse with the male; but concerning these topics we shall treat more precisely later on. In the case of certain fishes, however, after they have spontaneously generated eggs, these eggs develop into living animals; only that in certain of these cases development is spontaneous, and in others is not independent of the male; and the method of proceeding in regard to these matters will set forth by and by, for the method is somewhat like to the method followed in the case of birds. But whensoever creatures are spontaneously generated, either in other animals, in the soil, or on plants, or in the parts of these, and when such are generated male and female, then from the copulation of such spontaneously generated males and females there is generated a something-a something never identical in shape with the parents, but a something imperfect. For instance, the issue of copulation in lice is nits; in flies, grubs; in fleas, grubs egg-like in shape; and from these issues the parent-species is never reproduced, nor is any animal produced at all, but the like nondescripts only.

First, then, we must proceed to treat of 'covering' in regard to such animals as cover and are covered; and then after this to treat in due order of other matters, both the exceptional and those of general occurrence.

Part 2

Those animals, then, cover and are covered in which there is a duality of sex, and the modes of covering in such animals are not in all cases similar nor analogous. For the red-blooded animals that are viviparous and furnished with feet have in all cases organs adapted for procreation, but the sexes do not in all cases come together in like manner. Thus, opisthuretic animals copulate with a rear-ward presentment, as is the case with the lion, the hare, and the lynx; though, by the way, in the case of the hare, the female is often observed to cover the male.

The case is similar in most other such animals; that is to say, the majority of quadrupeds copulate as best they can, the male mounting the female; and this is the only method of copulating adopted by birds, though there are certain diversities of method observed even in birds. For in some cases the female squats on the ground and the male mounts on top of her, as is the case with the cock and hen bustard, and the barn-door cock and hen; in other cases, the male mounts without the female squatting, as with the male and female crane; for, with these birds, the male mounts on to the back of the female and covers her, and like the cock-sparrow consumes but very little time in the operation. Of quadrupeds, bears perform the operation lying prone on one another, in the same way as other quadrupeds do while standing up; that is to say, with the belly of the male pressed to the back of the female. Hedgehogs copulate erect, belly to belly.

With regard to large-sized vivipara, the hind only very rarely sustains the mounting of the stag to the full conclusion of the operation, and the same is the case with the cow as regards the bull, owing to the rigidity of the penis of the bull. In point of fact, the females of these animals elicit the sperm of the male in the act of withdrawing from underneath him; and, by the way, this phenomenon has been observed in the case of the stag and hind, domesticated, of course. Covering with the wolf is the same as with the dog. Cats do not copulate with a rearward presentment on the part of the female, but the male stands erect and the female puts herself underneath him; and, by the way, the female cat is peculiarly lecherous, and wheedles the male on to sexual commerce, and caterwauls during the operation. Camels copulate with the female in a sitting posture, and the male straddles over and covers her, not with the hinder presentment on the female's part but like the other quadrupeds mentioned above, and they pass the whole day long in the operation; when thus engaged they retire to lonely spots, and none but their keeper dare approach them. And, be it observed, the penis of the camel is so sinewy that bow-strings are manufactured out of it. Elephants, also, copulate in lonely places, and especially by river-sides in their usual haunts; the female squats down, and straddles with her legs, and the male mounts and covers her. The seal covers like all opisthuretic animals, and in this species the copulation extends over a lengthened time, as is the case with the dog and bitch; and the penis in the male seal is exceptionally large.

Part 3

Oviparous quadrupeds cover one another in the same way. That is to say, in some cases the male mounts the female precisely as in the viviparous animals, as is observed in both the land and the sea tortoise....And these creatures have an organ in which the ducts converge, and with which they perform the act of copulation, as is also observed in the toad, the frog, and all other animals of the same group.

Part 4

Long animals devoid of feet, like serpents and muraenae, intertwine in coition, belly to belly. And, in fact, serpents coil round one another so tightly as to present the appearance of a single serpent with a pair of heads. The same mode is followed by the saurians; that is to say, they coil round one another in the act of coition.

Part 5

All fishes, with the exception of the flat selachians, lie down side by side, and copulate belly to belly. Fishes, however, that are flat and furnished with tails-as the ray, the trygon, and the like-copulate not only in this way, but also, where the tail from its thinness is no impediment, by mounting of the male upon the female, belly to back. But the rhina or angel-fish, and other like fishes where the tail is large, copulate only by rubbing against one another sideways, belly to belly. Some men assure us that they have seen some of the selachia copulating hindways, dog and bitch. In the cartilaginous species the female is larger than the male; and the same is the case with other fishes for the most part. And among cartilaginous fishes are included, besides those already named, the bos, the lamia, the aetos, the narce or torpedo, the fishing-frog, and all the galeodes or sharks and dogfish. Cartilaginous fishes, then, of all kinds, have in many instances been observed copulating in the way above mentioned; for, by the way, in viviparous animals the process of copulation is of longer duration than in the ovipara.

It is the same with the dolphin and with all cetaceans; that is to say, they come side by side, male and female, and copulate, and the act extends over a time which is neither short nor very long.

Again, in cartilaginous fishes the male, in some species, differs from the female in the fact that he is furnished with two appendages hanging down from about the exit of the residuum, and that the female is not so furnished; and this distinction between the sexes is observed in all the species of the sharks and dog-fish.

Now neither fishes nor any animals devoid of feet are furnished with testicles, but male serpents and male fishes have a pair of ducts which fill with milt or sperm at the rutting season, and discharge, in all cases, a milk-like juice. These ducts unite, as in birds; for birds, by the way, have their testicles in their interior, and so have all ovipara that are furnished with feet. And this union of the ducts is so far continued and of such extension as to enter the receptive organ in the female.

In viviparous animals furnished with feet there is outwardly one and the same duct for the sperm and the liquid residuum; but there are separate ducts internally, as has been observed in the differentiation of the organs. And with such animals as are not viviparous the same passage serves for the discharge also of the solid residuum; although, internally, there are two passages, separate but near to one another. And these remarks apply to both male and female; for these animals are unprovided with a bladder except in the case of the tortoise; and the she-tortoise, though furnished with a bladder, has only one passage; and tortoises, by the way, belong to the ovipara.

In the case of oviparous fishes the process of coition is less open to observation. In point of fact, some are led by the want of actual observation to surmise that the female becomes impregnated by swallowing the seminal fluid of the male. And there can be no doubt that this proceeding on the part of the female is often witnessed; for at the rutting season the females follow the males and perform this operation, and strike the males with their mouths under the belly, and the males are thereby induced to part with the sperm sooner and more plentifully. And, further, at the spawning season the males go in pursuit of the females, and, as the female spawns, the males swallow the eggs; and the species is continued in existence by the spawn that survives this process. On the coast of Phoenicia they take advantage of these instinctive propensities of the two sexes to catch both one and the other: that is to say, by using the male of the grey mullet as a decoy they collect and net the female, and by using the female, the male.

The repeated observation of this phenomenon has led to the notion that the process was equivalent to coition, but the fact is that a similar phenomenon is observable in quadrupeds. For at the rutting seasons both the males and the females take to running at their genitals, and the two sexes take to smelling each other at those parts. (With partridges, by the way, if the female gets to leeward of the male, she becomes thereby impregnated. And often when they happen to be in heat she is affected in this wise by the voice of the male, or by his breathing down on her as he flies overhead; and, by the way, both the male and the female partridge keep the mouth wide open and protrude the tongue in the process of coition.)

The actual process of copulation on the part of oviparous fishes is seldom accurately observed, owing to the fact that they very soon fall aside and slip asunder. But, for all that, the process has been observed to take place in the manner above described.

Part 6

Molluscs, such as the octopus, the sepia, and the calamari, have sexual intercourse all in the same way; that is to say, they unite at the mouth, by an interlacing of their tentacles. When, then, the octopus rests its so-called head against the ground and spreads abroad its tentacles, the other sex fits into the outspreading of these tentacles, and the two sexes then bring their suckers into mutual connexion.

Some assert that the male has a kind of penis in one of his tentacles, the one in which are the largest suckers; and they further assert that the organ is tendinous in character, growing attached right up to the middle of the tentacle, and that the latter enables it to enter the nostril or funnel of the female.

Now cuttle-fish and calamaries swim about closely intertwined, with mouths and tentacles facing one another and fitting closely together, and swim thus in opposite directions; and they fit their so-called nostrils into one another, and the one sex swims backwards and the other frontwards during the operation. And the female lays its spawn by the so-called 'blow-hole'; and, by the way, some declare that it is at this organ that the coition really takes place.

Part 7

Crustaceans copulate, as the crawfish, the lobster, the carid and the like, just like the opisthuretic quadrupeds, when the one animal turns up its tail and the other puts his tail on the other's tail. Copulation takes place in the early spring, near to the shore; and, in fact, the process has often been observed in the case of all these animals. Sometimes it takes place about the time when the figs begin to ripen. Lobsters and carids copulate in like manner.

Crabs copulate at the front parts of one another, belly to belly, throwing their overlapping opercula to meet one another: first the smaller crab mounts the larger at the rear; after he has mounted, the larger one turns on one side. Now, the female differs in no respect from the male except in the circumstance that its operculum is larger, more elevated, and more hairy, and into this operculum it spawns its eggs and in the same neighbourhood is the outlet of the residuum. In the copulative process of these animals there is no protrusion of a member from one animal into the other.

Part 8

Insects copulate at the hinder end, and the smaller individuals mount the larger; and the smaller individual is I I is the male. The female pushes from underneath her sexual organ into the body of the male above, this being the reverse of the operation observed in other creatures; and this organ in the case of some insects appears to be disproportionately large when compared to the size of the body, and that too in very minute creatures; in some insects the disproportion is not so striking. This phenomenon may be witnessed if anyone will pull asunder flies that are copulating; and, by the way, these creatures are, under the circumstances, averse to separation; for the intercourse of the sexes in their case is of long duration, as may be observed with common everyday insects, such as the fly and the cantharis. They all copulate in the manner above described, the fly, the cantharis, the sphondyle, (the phalangium spider) any others of the kind that copulate at all. The phalangia-that is to say, such of the species as spin webs-perform the operation in the following way: the female takes hold of the suspended web at the middle and gives a pull, and the male gives a counter pull; this operation they repeat until they are drawn in together and interlaced at the hinder ends; for, by the way, this mode of copulation suits them in consequence of the rotundity of their stomachs.

So much for the modes of sexual intercourse in all animals; but, with regard to the same phenomenon, there are definite laws followed as regards the season of the year and the age of the animal.

Animals in general seem naturally disposed to this intercourse at about the same period of the year, and that is when winter is changing into summer. And this is the season of spring, in which almost all things that fly or walk or swim take to pairing. Some animals pair and breed in autumn also and in winter, as is the case with certain aquatic animals and certain birds. Man pairs and breeds at all seasons, as is the case also with domesticated animals, owing to the shelter and good feeding they enjoy: that is to say, with those whose period of gestation is also comparatively brief, as the sow and the bitch, and with those birds that breed frequently. Many animals time the season of intercourse with a view to the right nurture subsequently of their young. In the human species, the male is more under sexual excitement in winter, and the female in summer.

With birds the far greater part, as has been said, pair and breed during the spring and early summer, with the exception of the halcyon.

The halcyon breeds at the season of the winter solstice. Accordingly, when this season is marked with calm weather, the name of 'halcyon days' is given to the seven days preceding, and to as many following, the solstice; as Simonides the poet says:

God lulls for fourteen days the winds to sleep In winter; and this temperate interlude Men call the Holy Season, when the deep Cradles the mother Halcyon and her brood.

And these days are calm, when southerly winds prevail at the solstice, northerly ones having been the accompaniment of the Pleiades. The halcyon is said to take seven days for building her nest, and the other seven for laying and hatching her eggs. In our country there are not always halcyon days about the time of the winter solstice, but in the Sicilian seas this season of calm is almost periodical. The bird lays about five eggs.

Part 9

(The aithyia, or diver, and the larus, or gull, lay their eggs on rocks bordering on the sea, two or three at a time; but the gull lays in the summer, and the diver at the beginning of spring, just after the winter solstice, and it broods over its eggs as birds do in general. And neither of these birds resorts to a hiding-place.)

The halcyon is the most rarely seen of all birds. It is seen only about the time of the setting of the Pleiades and the winter solstice. When ships are lying at anchor in the roads, it will hover about a vessel and then disappear in a moment, and Stesichorus in one of his poems alludes to this peculiarity. The nightingale also breeds at the beginning of summer, and lays five or six eggs; from autumn until spring it retires to a hiding-place.

Insects copulate and breed in winter also, that is when the weather is fine and south winds prevail; such, I mean, as do not hibernate, as the fly and the ant. The greater part of wild animals bring forth once and once only in the year, except in the case of animals like the hare, where the female can become superfetally impregnated.

In like manner the great majority of fishes breed only once a year, like the shoal-fishes (or, in other words, such as are caught in nets), the tunny, the pelamys, the grey mullet, the chalcis, the mackerel, the sciaena, the psetta and the like, with the exception of the labrax or bass; for this fish (alone amongst those mentioned) breeds twice a year, and the second brood is the weaker of the two. The trichias and the rock-fishes breed twice a year; the red mullet breeds thrice a year, and is exceptional in this respect. This conclusion in regard to the red mullet is inferred from the spawn; for the spawn of the fish may be seen in certain places at three different times of the year.

The scorpaena breeds twice a year. The sargue breeds twice, in the spring and in the autumn. The saupe breeds once a year only, in the autumn. The female tunny breeds only once a year, but owing to the fact that the fish in some cases spawn early and in others late, it looks as though the fish bred twice over. The first spawning takes place in December before the solstice, and the latter spawning in the spring. The male tunny differs from the female in being unprovided with the fin beneath the belly which is called aphareus.

Part 10

Of cartilaginous fishes, the rhina or angelfish is the only one that breeds twice; for it breeds at the beginning of autumn, and at the setting of the Pleiads: and, of the two seasons, it is in better condition in the autumn. It engenders at a birth seven or eight young. Certain of the dog-fishes, for example the spotted dog, seem to breed twice a month, and this results from the circumstance that the eggs do not all reach maturity at the same time.

Some fishes breed at all seasons, as the muraena. This animal lays a great number of eggs at a time; and the young when hatched are very small but grow with great rapidity, like the young of the hippurus, for these fishes from being diminutive at the outset grow with exceptional rapidity to an exceptional size. (Be it observed that the muraena breeds at all seasons, but the hippurus only in the spring. The smyrus differs from the smyraena; for the muraena is mottled and weakly, whereas the smyrus is strong and of one uniform colour, and the colour resembles that of the pine-tree, and the animal has teeth inside and out. They say that in this case, as in other similar ones, the one is the male, and the other the female, of a single species. They come out on to the land, and are frequently caught.) Fishes, then, as a general rule, attain their full growth with great rapidity, but this is especially the case, among small fishes, with the coracine or crow-fish: it spawns, by the way, near the shore, in weedy and tangled spots. The orphus also, or sea-perch, is small at first, and rapidly attains a great size. The pelamys and the tunny breed in the Euxine, and nowhere else. The cestreus or mullet, the chrysophrys or gilt-head, and the labrax or basse, breed best where rivers run into the sea. The orcys or large-sized tunny, the scorpis, and many other species spawn in the open sea.

Part 11

Fish for the most part breed some time or other during the three months between the middle of March and the middle of June. Some few breed in autumn: as, for instance, the saupe and the sargus, and such others of this sort as breed shortly before the autumn equinox; likewise the electric ray and the angel-fish. Other fishes breed both in winter and in summer, as was previously observed: as, for instance, in winter-time the bass, the grey mullet, and the belone or pipe-fish; and in summer-time, from the middle of June to the middle of July, the female tunny, about the time of the summer solstice; and the tunny lays a sac-like enclosure in which are contained a number of small eggs. The ryades or shoal-fishes breed in summer.

Of the grey mullets, the chelon begins to be in roe between the middle of November and the middle of December; as also the sargue, and the smyxon or myxon, and the cephalus; and their period of gestation is thirty days. And, by the way, some of the grey mullet species are not produced from copulation, but grow spontaneously from mud and sand.

As a general rule, then, fishes are in roe in the spring-time; while some, as has been said, are so in summer, in autumn, or in winter. But whereas the impregnation in the spring-time follows a general law, impregnation in the other seasons does not follow the same rule either throughout or within the limits of one genus; and, further, conception in these variant seasons is not so prolific. And, indeed, we must bear this in mind, that just as with plants and quadrupeds diversity of locality has much to do not only with general physical health but also with the comparative frequency of sexual intercourse and generation, so also with regard to fishes locality of itself has much to do not only in regard to the size and vigour of the creature, but also in regard to its parturition and its copulations, causing the same species to breed more often in one place and more seldom in another.

Part 12

The molluscs also breed in spring. Of the marine molluscs one of the first to breed is the sepia. It spawns at all times of the day and its period of gestation is fifteen days. After the female has laid her eggs, the male comes and discharges the milt over the eggs, and the eggs thereupon harden. And the two sexes of this animal go about in pairs, side by side; and the male is more mottled and more black on the back than the female.

The octopus pairs in winter and breeds in spring, lying hidden for about two months. Its spawn is shaped like a vine-tendril, and resembles the fruit of the white poplar; the creature is extraordinarily prolific, for the number of individuals that come from the spawn is something incalculable. The male differs from the female in the fact that its head is longer, and that the organ called by the fishermen its penis, in the tentacle, is white. The female, after laying her eggs, broods over them, and in consequence gets out of condition, by reason of not going in quest of food during the hatching period.

The purple murex breeds about springtime, and the ceryx at the close of the winter. And, as a general rule, the testaceans are found to be furnished with their so-called eggs in spring-time and in autumn, with the exception of the edible urchin; for this animal has the so-called eggs in most abundance in these seasons, but at no season is unfurnished with them; and it is furnished with them in especial abundance in warm weather or when a full moon is in the sky. Only, by the way, these remarks do not apply to the sea-urchin found in the Pyrrhaean Straits, for this urchin is at its best for table purposes in the winter; and these urchins are small but full of eggs.

Snails are found by observations to become in all cases impregnated about the same season.

Part 13

(Of birds the wild species, as has been stated, as a general rule pair and breed only once a year. The swallow, however, and the blackbird breed twice. With regard to the blackbird, however, its first brood is killed by inclemency of weather (for it is the earliest of all birds to breed), but the second brood it usually succeeds in rearing.

Birds that are domesticated or that are capable of domestication breed frequently, just as the common pigeon breeds all through the summer, and as is seen in the barn-door hen; for the barn-door cock and hen have intercourse, and the hen breeds, at all seasons alike: excepting by the way, during the days about the winter solstice.

Of the pigeon family there are many diversities; for the peristera or common pigeon is not identical with the peleias or rock-pigeon. In other words, the rock-pigeon is smaller than the common pigeon, and is less easily domesticated; it is also black, and small, red-footed and rough-footed; and in consequence of these peculiarities it is neglected by the pigeon-fancier. The largest of all the pigeon species is the phatta or ring-dove; and the next in size is the oenas or stock-dove; and the stock-dove is a little larger than the common pigeon. The smallest of all the species is the turtle-dove. Pigeons breed and hatch at all seasons, if they are furnished with a sunny place and all requisites; unless they are so furnished, they breed only in the summer. The spring brood is the best, or the autumn brood. At all events, without doubt, the produce of the hot season, the summer brood, is the poorest of the three.)

Part 14

Further, animals differ from one another in regard to the time of life that is best adapted for sexual intercourse.

To begin with, in most animals the secretion of the seminal fluid and its generative capacity are not phenomena simultaneously manifested, but manifested successively. Thus, in all animals, the earliest secretion of sperm is unfruitful, or if it be fruitful the issue is comparatively poor and small. And this phenomenon is especially observable in man, in viviparous quadrupeds, and in birds; for in the case of man and the quadruped the offspring is smaller, and in the case of the bird, the egg.

For animals that copulate, of one and the same species, the age for maturity is in most species tolerably uniform, unless it occurs prematurely by reason of abnormality, or is postponed by physical injury.

In man, then, maturity is indicated by a change of the tone of voice, by an increase in size and an alteration in appearance of the sexual organs, as also in an increase of size and alteration in appearance of the breasts; and above all, in the hair-growth at the pubes. Man begins to possess seminal fluid about the age of fourteen, and becomes generatively capable at about the age of twenty-one years.

In other animals there is no hair-growth at the pubes (for some animals have no hair at all, and others have none on the belly, or less on the belly than on the back), but still, in some animals the change of voice is quite obvious; and in some animals other organs give indication of the commencing secretion of the sperm and the onset of generative capacity. As a general rule the female is sharper-toned in voice than the male, and the young animal than the elder; for, by the way, the stag has a much deeper-toned bay than the hind. Moreover, the male cries chiefly at rutting time, and the female under terror and alarm; and the cry of the female is short, and that of the male prolonged. With dogs also, as they grow old, the tone of the bark gets deeper.

There is a difference observable also in the neighings of horses. That is to say, the female foal has a thin small neigh, and the male foal a small neigh, yet bigger and deeper-toned than that of the female, and a louder one as time goes on. And when the young male and female are two years old and take to breeding, the neighing of the stallion becomes loud and deep, and that of the mare louder and shriller than heretofore; and this change goes on until they reach the age of about twenty years; and after this time the neighing in both sexes becomes weaker and weaker.

As a rule, then, as was stated, the voice of the male differs from the voice of the female, in animals where the voice admits of a continuous and prolonged sound, in the fact that the note in the male voice is more deep and bass; not, however, in all animals, for the contrary holds good in the case of some, as for instance in kine: for here the cow has a deeper note than the bull, and the calves a deeper note than the cattle. And we can thus understand the change of voice in animals that undergo gelding; for male animals that undergo this process assume the characters of the female.

The following are the ages at which various animals become capacitated for sexual commerce. The ewe and the she-goat are sexually mature when one year old, and this statement is made more confidently in respect to the she-goat than to the ewe; the ram and the he-goat are sexually mature at the same age. The progeny of very young individuals among these animals differs from that of other males: for the males improve in the course of the second year, when they become fully mature. The boar and the sow are capable of intercourse when eight months old, and the female brings forth when one year old, the difference corresponding to her period of gestation. The boar is capable of generation when eight months old, but, with a sire under a year in age, the litter is apt to be a poor one. The ages, however, are not invariable; now and then the boar and the sow are capable of intercourse when four months old, and are capable of producing a litter which can be reared when six months old; but at times the boar begins to be capable of intercourse when ten months. He continues sexually mature until he is three years old. The dog and the bitch are, as a rule, sexually capable and sexually receptive when a year old, and sometimes when eight months old; but the priority in date is more common with the dog than with the bitch. The period of gestation with the bitch is sixty days, or sixty-one, or sixty-two, or sixty-three at the utmost; the period is never under sixty days, or, if it is, the litter comes to no good. The bitch, after delivering a litter, submits to the male in six months, but not before. The horse and the mare are, at the earliest, sexually capable and sexually mature when two years old; the issue, however, of parents of this age is small and poor. As a general rule these animals are sexually capable when three years old, and they grow better for breeding purposes until they reach twenty years. The stallion is sexually capable up to the age of thirty-three years, and the mare up to forty, so that, in point of fact, the animals are sexually capable all their lives long; for the stallion, as a rule, lives for about thirty-five years, and the mare for a little over forty; although, by the way, a horse has known to live to the age of seventy-five. The ass and the she-ass are sexually capable when thirty months old; but, as a rule, they are not generatively mature until they are three years old, or three years and a half. An instance has been known of a she-ass bearing and bringing forth a foal when only a year old. A cow has been known to calve when only a year old, and the calf grew as big as might be expected, but no more. So much for the dates in time at which these animals attain to generative capacity.

In the human species, the male is generative, at the longest, up to seventy years, and the female up to fifty; but such extended periods are rare. As a rule, the male is generative up to the age of sixty-five, and to the age of forty-five the female is capable of conception.

The ewe bears up to eight years, and, if she be carefully tended, up to eleven years; in fact, the ram and the ewe are sexually capable pretty well all their lives long. He-goats, if they be fat, are more or less unserviceable for breeding; and this, by the way, is the reason why country folk say of a vine when it stops bearing that it is 'running the goat'. However, if an over-fat he-goat be thinned down, he becomes sexually capable and generative.

Rams single out the oldest ewes for copulation, and show no regard for the young ones. And, as has been stated, the issue of the younger ewes is poorer than that of the older ones.

The boar is good for breeding purposes until he is three years of age; but after that age his issue deteriorates, for after that age his vigour is on the decline. The boar is most capable after a good feed, and with the first sow it mounts; if poorly fed or put to many females, the copulation is abbreviated, and the litter is comparatively poor. The first litter of the sow is the fewest in number; at the second litter she is at her prime. The animal, as it grows old, continues to breed, but the sexual desire abates. When they reach fifteen years, they become unproductive, and are getting old. If a sow be highly fed, it is all the more eager for sexual commerce, whether old or young; but, if it be over-fattened in pregnancy, it gives the less milk after parturition. With regard to the age of the parents, the litter is the best when they are in their prime; but with regard to the seasons of the year, the litter is the best that comes at the beginning of winter; and the summer litter the poorest, consisting as it usually does of animals small and thin and flaccid. The boar, if it be well fed, is sexually capable at all hours, night and day; but otherwise is peculiarly salacious early in the morning. As it grows old the sexual passion dies away, as we have already remarked. Very often a boar, when more or less impotent from age or debility, finding itself unable to accomplish the sexual commerce with due speed, and growing fatigued with the standing posture, will roll the sow over on the ground, and the pair will conclude the operation side by side of one another. The sow is sure of conception if it drops its lugs in rutting time; if the ears do not thus drop, it may have to rut a second time before impregnation takes place.

Bitches do not submit to the male throughout their lives, but only until they reach a certain maturity of years. As a general rule, they are sexually receptive and conceptive until they are twelve years old; although, by the way, cases have been known where dogs and bitches have been respectively procreative and conceptive to the ages of eighteen and even of twenty years. But, as a rule, age diminishes the capability of generation and of conception with these animals as with all others.

The female of the camel is opisthuretic, and submits to the male in the way above described; and the season for copulation in Arabia is about the month of October. Its period of gestation is twelve months; and it is never delivered of more than one foal at a time. The female becomes sexually receptive and the male sexually capable at the age of three years. After parturition, an interval of a year elapses before the female is again receptive to the male.

The female elephant becomes sexually receptive when ten years old at the youngest, and when fifteen at the oldest; and the male is sexually capable when five years old, or six. The season for intercourse is spring. The male allows an interval of three years to elapse after commerce with a female: and, after it has once impregnated a female, it has no intercourse with her again. The period of gestation with the female is two years; and only one young animal is produced at a time, in other words it is uniparous. And the embryo is the size of a calf two or three months old.

Part 15

So much for the copulations of such animals as copulate. We now proceed to treat of generation both with respect to copulating and non-copulating animals, and we shall commence with discussing the subject of generation in the case of the testaceans.

The testacean is almost the only genus that throughout all its species is non-copulative.

The porphyrae, or purple murices, gather together to some one place in the spring-time, and deposit the so-called 'honeycomb'. This substance resembles the comb, only that it is not so neat and delicate; and looks as though a number of husks of white chick-peas were all stuck together. But none of these structures has any open passage, and the porphyra does not grow out of them, but these and all other testaceans grow out of mud and decaying matter. The substance, is, in fact, an excretion of the porphyra and the ceryx; for it is deposited by the ceryx as well. Such, then, of the testaceans as deposit the honeycomb are generated spontaneously like all other testaceans, but they certainly come in greater abundance in places where their congeners have been living previously. At the commencement of the process of depositing the honeycomb, they throw off a slippery mucus, and of this the husklike formations are composed. These formations, then, all melt and deposit their contents on the ground, and at this spot there are found on the ground a number of minute porphyrae, and porphyrae are caught at times with these animalculae upon them, some of which are too small to be differentiated in form. If the porphyrae are caught before producing this honey-comb, they sometimes go through the process in fishing-creels, not here and there in the baskets, but gathering to some one spot all together, just as they do in the sea; and owing to the narrowness of their new quarters they cluster together like a bunch of grapes.

There are many species of the purple murex; and some are large, as those found off Sigeum and Lectum; others are small, as those found in the Euripus, and on the coast of Caria. And those that are found in bays are large and rough; in most of them the peculiar bloom from which their name is derived is dark to blackness, in others it is reddish and small in size; some of the large ones weigh upwards of a mina apiece. But the specimens that are found along the coast and on the rocks are small-sized, and the bloom in their case is of a reddish hue. Further, as a general rule, in northern waters the bloom is blackish, and in southern waters of a reddish hue. The murex is caught in the spring-time when engaged in the construction of the honeycomb; but it is not caught at any time about the rising of the dog-star, for at that period it does not feed, but conceals itself and burrows. The bloom of the animal is situated between the mecon (or quasi-liver) and the neck, and the co-attachment of these is an intimate one. In colour it looks like a white membrane, and this is what people extract; and if it be removed and squeezed it stains your hand with the colour of the bloom. There is a kind of vein that runs through it, and this quasi-vein would appear to be in itself the bloom. And the qualities, by the way, of this organ are astringent. It is after the murex has constructed the honeycomb that the bloom is at its worst. Small specimens they break in pieces, shells and all, for it is no easy matter to extract the organ; but in dealing with the larger ones they first strip off the shell and then abstract the bloom. For this purpose the neck and mecon are separated, for the bloom lies in between them, above the so-called stomach; hence the necessity of separating them in abstracting the bloom. Fishermen are anxious always to break the animal in pieces while it is yet alive, for, if it die before the process is completed, it vomits out the bloom; and for this reason the fishermen keep the animals in creels, until they have collected a sufficient number and can attend to them at their leisure. Fishermen in past times used not to lower creels or attach them to the bait, so that very often the animal got dropped off in the pulling up; at present, however, they always attach a basket, so that if the animal fall off it is not lost. The animal is more inclined to slip off the bait if it be full inside; if it be empty it is difficult to shake it off. Such are the phenomena connected with the porphyra or murex.

The same phenomena are manifested by the ceryx or trumpet-shell; and the seasons are the same in which the phenomena are observable. Both animals, also, the murex and the ceryx, have their opercula similarly situated-and, in fact, all the stromboids, and this is congenital with them all; and they feed by protruding the so-called tongue underneath the operculum. The tongue of the murex is bigger than one's finger, and by means of it, it feeds, and perforates conchylia and the shells of its own kind. Both the murex and the ceryx are long lived. The murex lives for about six years; and the yearly increase is indicated by a distinct interval in the spiral convolution of the shell.

The mussel also constructs a honeycomb. With regard to the limnostreae, or lagoon oysters, wherever you have slimy mud there you are sure to find them beginning to grow. Cockles and clams and razor-fishes and scallops row spontaneously in sandy places. The pinna grows straight up from its tuft of anchoring fibres in sandy and slimy places; these creatures have inside them a parasite nicknamed the pinna-guard, in some cases a small carid and in other cases a little crab; if the pinna be deprived of this pinna-guard it soon dies.

As a general rule, then, all testaceans grow by spontaneous generation in mud, differing from one another according to the differences of the material; oysters growing in slime, and cockles and the other testaceans above mentioned on sandy bottoms; and in the hollows of the rocks the ascidian and the barnacle, and common sorts, such as the limpet and the nerites. All these animals grow with great rapidity, especially the murex and the scallop; for the murex and the scallop attain their full growth in a year. In some of the testaceans white crabs are found, very diminutive in size; they are most numerous in the trough shaped mussel. In the pinna also is found the so-called pinna-guard. They are found also in the scallop and in the oyster; these parasites never appear to grow in size. Fishermen declare that the parasite is congenital with the larger animal. (Scallops burrow for a time in the sand, like the murex.)

(Shell-fish, then, grow in the way above mentioned; and some of them grow in shallow water, some on the sea-shore, some in rocky places, some on hard and stony ground, and some in sandy places.) Some shift about from place to place, others remain permanent on one spot. Of those that keep to one spot the pinnae are rooted to the ground; the razor-fish and the clam keep to the same locality, but are not so rooted; but still, if forcibly removed they die.

(The star-fish is naturally so warm that whatever it lays hold of is found, when suddenly taken away from the animal, to have undergone a process like boiling. Fishermen say that the star-fish is a great pest in the Strait of Pyrrha. In shape it resembles a star as seen in an ordinary drawing. The so-called 'lungs' are generated spontaneously. The shells that painters use are a good deal thicker, and the bloom is outside the shell on the surface. These creatures are mostly found on the coast of Caria.)

The hermit-crab grows spontaneously out of soil and slime, and finds its way into untenanted shells. As it grows it shifts to a larger shell, as for instance into the shell of the nerites, or of the strombus or the like, and very often into the shell of the small ceryx. After entering new shell, it carries it about, and begins again to feed, and, by and by, as it grows, it shifts again into another larger one.

Part 16

Moreover, the animals that are unfurnished with shells grow spontaneously, like the testaceans, as, for instance, the sea-nettles and the sponges in rocky caves.

Of the sea-nettle, or sea-anemone, there are two species; and of these one species lives in hollows and never loosens its hold upon the rocks, and the other lives on smooth flat reefs, free and detached, and shifts its position from time to time. (Limpets also detach themselves, and shift from place to place.)

In the chambered cavities of sponges pinna-guards or parasites are found. And over the chambers there is a kind of spider's web, by the opening and closing of which they catch mute fishes; that is to say, they open the web to let the fish get in, and close it again to entrap them.

Of sponges there are three species; the first is of loose porous texture, the second is close textured, the third, which is nicknamed 'the sponge of Achilles', is exceptionally fine and close-textured and strong. This sponge is used as a lining to helmets and greaves, for the purpose of deadening the sound of the blow; and this is a very scarce species. Of the close textured sponges such as are particularly hard and rough are nicknamed 'goats'.

Sponges grow spontaneously either attached to a rock or on sea-beaches, and they get their nutriment in slime: a proof of this statement is the fact that when they are first secured they are found to be full of slime. This is characteristic of all living creatures that get their nutriment by close local attachment. And, by the way, the close-textured sponges are weaker than the more openly porous ones because their attachment extends over a smaller area.

It is said that the sponge is sensitive; and as a proof of this statement they say that if the sponge is made aware of an attempt being made to pluck it from its place of attachment it draws itself together, and it becomes a difficult task to detach it. It makes a similar contractile movement in windy and boisterous weather, obviously with the object of tightening its hold. Some persons express doubts as to the truth of this assertion; as, for instance, the people of Torone.

The sponge breeds parasites, worms, and other creatures, on which, if they be detached, the rock-fishes prey, as they prey also on the remaining stumps of the sponge; but, if the sponge be broken off, it grows again from the remaining stump and the place is soon as well covered as before.

The largest of all sponges are the loose-textured ones, and these are peculiarly abundant on the coast of Lycia. The softest are the close-textured sponges; for, by the way, the so-called sponges of Achilles are harder than these. As a general rule, sponges that are found in deep calm waters are the softest; for usually windy and stormy weather has a tendency to harden them (as it has to harden all similar growing things), and to arrest their growth. And this accounts for the fact that the sponges found in the Hellespont are rough and close-textured; and, as a general rule, sponges found beyond or inside Cape Malea are, respectively, comparatively soft or comparatively hard. But, by the way, the habitat of the sponge should not be too sheltered and warm, for it has a tendency to decay, like all similar vegetable-like growths. And this accounts for the fact that the sponge is at its best when found in deep water close to shore; for owing to the depth of the water they enjoy shelter alike from stormy winds and from excessive heat.

Whilst they are still alive and before they are washed and cleaned, they are blackish in colour. Their attachment is not made at one particular spot, nor is it made all over their bodies; for vacant pore-spaces intervene. There is a kind of membrane stretched over the under parts; and in the under parts the points of attachment are the more numerous. On the top most of the pores are closed, but four or five are open and visible; and we are told by some that it is through these pores that the animal takes its food.

There is a particular species that is named the 'aplysia' or the 'unwashable', from the circumstance that it cannot be cleaned. This species has the large open and visible pores, but all the rest of the body is close-textured; and, if it be dissected, it is found to be closer and more glutinous than the ordinary sponge, and, in a word, something lung like in consistency. And, on all hands, it is allowed that this species is sensitive and long-lived. They are distinguished in the sea from ordinary sponges from the circumstance that the ordinary sponges are white while the slime is in them, but that these sponges are under any circumstances black.

And so much with regard to sponges and to generation in the testaceans.

Part 17

Of crustaceans, the female crawfish after copulation conceives and retains its eggs for about three months, from about the middle of May to about the middle of August; they then lay the eggs into the folds underneath the belly, and their eggs grow like grubs. This same phenomenon is observable in molluscs also, and in such fishes as are oviparous; for in all these cases the egg continues to grow.

The spawn of the crawfish is of a loose or granular consistency, and is divided into eight parts; for corresponding to each of the flaps on the side there is a gristly formation to which the spawn is attached, and the entire structure resembles a cluster of grapes; for each gristly formation is split into several parts. This is obvious enough if you draw the parts asunder; but at first sight the whole appears to be one and indivisible. And the largest are not those nearest to the outlet but those in the middle, and the farthest off are the smallest. The size of the small eggs is that of a small seed in a fig; and they are not quite close to the outlet, but placed middleways; for at both ends, tailwards and trunkwards, there are two intervals devoid of eggs; for it is thus that the flaps also grow. The side flaps, then, cannot close, but by placing the end flap on them the animal can close up all, and this end-flap serves them for a lid. And in the act of laying its eggs it seems to bring them towards the gristly formations by curving the flap of its tail, and then, squeezing the eggs towards the said gristly formations and maintaining a bent posture, it performs the act of laying. The gristly formations at these seasons increase in size and become receptive of the eggs; for the animal lays its eggs into these formations, just as the sepia lays its eggs among twigs and driftwood.

It lays its eggs, then, in this manner, and after hatching them for about twenty days it rids itself of them all in one solid lump, as is quite plain from outside. And out of these eggs crawfish form in about fifteen days, and these crawfish are caught at times less then a finger's breadth, or seventenths of an inch, in length. The animal, then, lays its eggs before the middle of September, and after the middle of that month throws off its eggs in a lump. With the humped carids or prawns the time for gestation is four months or thereabouts.

Crawfish are found in rough and rocky places, lobsters in smooth places, and neither crawfish nor lobsters are found in muddy ones; and this accounts for the fact that lobsters are found in the Hellespont and on the coast of Thasos, and crawfish in the neighbourhood of Sigeum and Mount Athos. Fishermen, accordingly, when they want to catch these various creatures out at sea, take bearings on the beach and elsewhere that tell them where the ground at the bottom is stony and where soft with slime. In winter and spring these animals keep in near to land, in summer they keep in deep water; thus at various times seeking respectively for warmth or coolness.

The so-called arctus or bear-crab lays its eggs at about the same time as the crawfish; and consequently in winter and in the spring-time, before laying their eggs, they are at their best, and after laying at their worst.

They cast their shell in the spring-time (just as serpents shed their so-called 'old-age' or slough), both directly after birth and in later life; this is true both of crabs and crawfish. And, by the way, all crawfish are long lived.

Part 18

Molluscs, after pairing and copulation, lay a white spawn; and this spawn, as in the case of the testacean, gets granular in time. The octopus discharges into its hole, or into a potsherd or into any similar cavity, a structure resembling the tendrils of a young vine or the fruit of the white poplar, as has been previously observed. The eggs, when the female has laid them, are clustered round the sides of the hole. They are so numerous that, if they be removed they suffice to fill a vessel much larger than the animal's body in which they were contained. Some fifty days later, the eggs burst and the little polypuses creep out, like little spiders, in great numbers; the characteristic form of their limbs is not yet to be discerned in detail, but their general outline is clear enough. And, by the way, they are so small and helpless that the greater number perish; it is a fact that they have been seen so extremely minute as to be absolutely without organization, but nevertheless when touched they moved. The eggs of the sepia look like big black myrtle-berries, and they are linked all together like a bunch of grapes, clustered round a centre, and are not easily sundered from one another: for the male exudes over them some moist glairy stuff, which constitutes the sticky gum. These eggs increase in size; and they are white at the outset, but black and larger after the sprinkling of the male seminal fluid.

When it has come into being the young sepia is first distinctly formed inside out of the white substance, and when the egg bursts it comes out. The inner part is formed as soon as the female lays the egg, something like a hail-stone; and out of this substance the young sepia grows by a head-attachment, just as young birds grow by a belly-attachment. What is the exact nature of the navel-attachment has not yet been observed, except that as the young sepia grows the white substance grows less and less in size, and at length, as happens with the yolk in the case of birds, the white substance in the case of the young sepia disappears. In the case of the young sepia, as in the case of the young of most animals, the eyes at first seem very large. To illustrate this by way of a figure, let A represent the ovum, B and C the eyes, and D the sepidium, or body of the little sepia. (See diagram.)

The female sepia goes pregnant in the spring-time, and lays its eggs after fifteen days of gestation; after the eggs are laid there comes in another fifteen days something like a bunch of grapes, and at the bursting of these the young sepiae issue forth. But if, when the young ones are fully formed, you sever the outer covering a moment too soon, the young creatures eject excrement, and their colour changes from white to red in their alarm.

Crustaceans, then, hatch their eggs by brooding over them as they carry them about beneath their bodies; but the octopus, the sepia, and the like hatch their eggs without stirring from the spot where they may have laid them, and this statement is particularly applicable to the sepia; in fact, the nest of the female sepia is often seen exposed to view close in to shore. The female octopus at times sits brooding over her eggs, and at other times squats in front of her hole, stretching out her tentacles on guard.

The sepia lays her spawn near to land in the neighbourhood of sea-weed or reeds or any off-sweepings such as brushwood, twigs, or stones; and fishermen place heaps of faggots here and there on purpose, and on to such heaps the female deposits a long continuous roe in shape like a vine tendril. It lays or spirts out the spawn with an effort, as though there were difficulty in the process. The female calamari spawns at sea; and it emits the spawn, as does the sepia, in the mass.

The calamari and the cuttle-fish are short-lived, as, with few exceptions, they never see the year out; and the same statement is applicable to the octopus.

From one single egg comes one single sepia; and this is likewise true of the young calamari.

The male calamari differs from the female; for if its gill-region be dilated and examined there are found two red formations resembling breasts, with which the male is unprovided. In the sepia, apart from this distinction in the sexes, the male, as has been stated, is more mottled than the female.

Part 19

With regard to insects, that the male is less than the female and that he mounts upon her back, and how he performs the act of copulation and the circumstance that he gives over reluctantly, all this has already been set forth, most cases of insect copulation this process is speedily followed up by parturition.

All insects engender grubs, with the exception of a species of butterfly; and the female of this species lays a hard egg, resembling the seed of the cnecus, with a juice inside it. But from the grub, the young animal does not grow out of a mere portion of it, as a young animal grows from a portion only of an egg, but the grub entire grows and the animal becomes differentiated out of it.

And of insects some are derived from insect congeners, as the venom-spider and the common-spider from the venom-spider and the common-spider, and so with the attelabus or locust, the acris or grasshopper, and the tettix or cicada. Other insects are not derived from living parentage, but are generated spontaneously: some out of dew falling on leaves, ordinarily in spring-time, but not seldom in winter when there has been a stretch of fair weather and southerly winds; others grow in decaying mud or dung; others in timber, green or dry; some in the hair of animals; some in the flesh of animals; some in excrements: and some from excrement after it has been voided, and some from excrement yet within the living animal, like the helminthes or intestinal worms. And of these intestinal worms there are three species: one named the flat-worm, another the round worm, and the third the ascarid. These intestinal worms do not in any case propagate their kind. The flat-worm, however, in an exceptional way, clings fast to the gut, and lays a thing like a melon-seed, by observing which indication the physician concludes that his patient is troubled with the worm.

The so-called psyche or butterfly is generated from caterpillars which grow on green leaves, chiefly leaves of the raphanus, which some call crambe or cabbage. At first it is less than a grain of millet; it then grows into a small grub; and in three days it is a tiny caterpillar. After this it grows on and on, and becomes quiescent and changes its shape, and is now called a chrysalis. The outer shell is hard, and the chrysalis moves if you touch it. It attaches itself by cobweb-like filaments, and is unfurnished with mouth or any other apparent organ. After a little while the outer covering bursts asunder, and out flies the winged creature that we call the psyche or butterfly. At first, when it is a caterpillar, it feeds and ejects excrement; but when it turns into the chrysalis it neither feeds nor ejects excrement.

The same remarks are applicable to all such insects as are developed out of the grub, both such grubs as are derived from the copulation of living animals and such as are generated without copulation on the part of parents. For the grub of the bee, the anthrena, and the wasp, whilst it is young, takes food and voids excrement; but when it has passed from the grub shape to its defined form and become what is termed a 'nympha', it ceases to take food and to void excrement, and remains tightly wrapped up and motionless until it has reached its full size, when it breaks the formation with which the cell is closed, and issues forth. The insects named the hypera and the penia are derived from similar caterpillars, which move in an undulatory way, progressing with one part and then pulling up the hinder parts by a bend of the body. The developed insect in each case takes its peculiar colour from the parent caterpillar.

From one particular large grub, which has as it were horns, and in other respects differs from grubs in general, there comes, by a metamorphosis of the grub, first a caterpillar, then the cocoon, then the necydalus; and the creature passes through all these transformations within six months. A class of women unwind and reel off the cocoons of these creatures, and afterwards weave a fabric with the threads thus unwound; a Coan woman of the name of Pamphila, daughter of Plateus, being credited with the first invention of the fabric. After the same fashion the carabus or stag-beetle comes from grubs that live in dry wood: at first the grub is motionless, but after a while the shell bursts and the stag-beetle issues forth.

From the cabbage is engendered the cabbageworm, and from the leek the prasocuris or leekbane; this creature is also winged. From the flat animalcule that skims over the surface of rivers comes the oestrus or gadfly; and this accounts for the fact that gadflies most abound in the neighbourhood of waters on whose surface these animalcules are observed. From a certain small, black and hairy caterpillar comes first a wingless glow-worm; and this creature again suffers a metamorphosis, and transforms into a winged insect named the bostrychus (or hair-curl).

Gnats grow from ascarids; and ascarids are engendered in the slime of wells, or in places where there is a deposit left by the draining off of water. This slime decays, and first turns white, then black, and finally blood-red; and at this stage there originate in it, as it were, little tiny bits of red weed, which at first wriggle about all clinging together, and finally break loose and swim in the water, and are hereupon known as ascarids. After a few days they stand straight up on the water motionless and hard, and by and by the husk breaks off and the gnats are seen sitting upon it, until the sun's heat or a puff of wind sets them in motion, when they fly away.

With all grubs and all animals that break out from the grub state, generation is due primarily to the heat of the sun or to wind.

Ascarids are more likely to be found, and grow with unusual rapidity, in places where there is a deposit of a mixed and heterogeneous kind, as in kitchens and in ploughed fields, for the contents of such places are disposed to rapid putrefaction. In autumn, also, owing to the drying up of moisture, they grow in unusual numbers.

The tick is generated from couch-grass. The cockchafer comes from a grub that is generated in the dung of the cow or the ass. The cantharus or scarabeus rolls a piece of dung into a ball, lies hidden within it during the winter, and gives birth therein to small grubs, from which grubs come new canthari. Certain winged insects also come from the grubs that are found in pulse, in the same fashion as in the cases described.

Flies grow from grubs in the dung that farmers have gathered up into heaps: for those who are engaged in this work assiduously gather up the compost, and this they technically term 'working-up' the manure. The grub is exceedingly minute to begin with; first even at this stage-it assumes a reddish colour, and then from a quiescent state it takes on the power of motion, as though born to it; it then becomes a small motionless grub; it then moves again, and again relapses into immobility; it then comes out a perfect fly, and moves away under the influence of the sun's heat or of a puff of air. The myops or horse-fly is engendered in timber. The orsodacna or budbane is a transformed grub; and this grub is engendered in cabbage-stalks. The cantharis comes from the caterpillars that are found on fig-trees or pear-trees or fir-trees--for on all these grubs are engendered-and also from caterpillars found on the dog-rose; and the cantharis takes eagerly to ill-scented substances, from the fact of its having been engendered in ill-scented woods. The conops comes from a grub that is engendered in the slime of vinegar.

And, by the way, living animals are found in substances that are usually supposed to be incapable of putrefaction; for instance, worms are found in long-lying snow; and snow of this description gets reddish in colour, and the grub that is engendered in it is red, as might have been expected, and it is also hairy. The grubs found in the snows of Media are large and white; and all such grubs are little disposed to motion. In Cyprus, in places where copper-ore is smelted, with heaps of the ore piled on day after day, an animal is engendered in the fire, somewhat larger than a blue bottle fly, furnished with wings, which can hop or crawl through the fire. And the grubs and these latter animals perish when you keep the one away from the fire and the other from the snow. Now the salamander is a clear case in point, to show us that animals do actually exist that fire cannot destroy; for this creature, so the story goes, not only walks through the fire but puts it out in doing so.

On the river Hypanis in the Cimmerian Bosphorus, about the time of the summer solstice, there are brought down towards the sea by the stream what look like little sacks rather bigger than grapes, out of which at their bursting issues a winged quadruped. The insect lives and flies about until the evening, but as the sun goes down it pines away, and dies at sunset having lived just one day, from which circumstance it is called the ephemeron.

As a rule, insects that come from caterpillars and grubs are held at first by filaments resembling the threads of a spider's web.

Such is the mode of generation of the insects above enumerated. but if the latter impregnation takes place during the change of the yellow

Part 20

The wasps that are nicknamed 'the ichneumons' (or hunters), less in size, by the way, than the ordinary wasp, kill spiders and carry off the dead bodies to a wall or some such place with a hole in it; this hole they smear over with mud and lay their grubs inside it, and from the grubs come the hunter-wasps. Some of the coleoptera and of the small and nameless insects make small holes or cells of mud on a wall or on a grave-stone, and there deposit their grubs.

With insects, as a general rule, the time of generation from its commencement to its completion comprises three or four weeks. With grubs and grub-like creatures the time is usually three weeks, and in the oviparous insects as a rule four. But, in the case of oviparous insects, the egg-formation comes at the close of seven days from copulation, and during the remaining three weeks the parent broods over and hatches its young; i.e. where this is the result of copulation, as in the case of the spider and its congeners. As a rule, the transformations take place in intervals of three or four days, corresponding to the lengths of interval at which the crises recur in intermittent fevers.

So much for the generation of insects. Their death is due to the shrivelling of their organs, just as the larger animals die of old age.

Winged insects die in autumn from the shrinking of their wings. The myops dies from dropsy in the eyes.

Part 21

With regard to the generation of bees different hypotheses are in vogue. Some affirm that bees neither copulate nor give birth to young, but that they fetch their young. And some say that they fetch their young from the flower of the callyntrum; others assert that they bring them from the flower of the reed, others, from the flower of the olive. And in respect to the olive theory, it is stated as a proof that, when the olive harvest is most abundant, the swarms are most numerous. Others declare that they fetch the brood of the drones from such things as above mentioned, but that the working bees are engendered by the rulers of the hive.

Now of these rulers there are two kinds: the better kind is red in colour, the inferior kind is black and variegated; the ruler is double the size of the working bee. These rulers have the abdomen or part below the waist half as large again, and they are called by some the 'mothers', from an idea that they bear or generate the bees; and, as a proof of this theory of their motherhood, they declare that the brood of the drones appears even when there is no ruler-bee in the hive, but that the bees do not appear in his absence. Others, again, assert that these insects copulate, and that the drones are male and the bees female.

The ordinary bee is generated in the cells of the comb, but the ruler-bees in cells down below attached to the comb, suspended from it, apart from the rest, six or seven in number, and growing in a way quite different from the mode of growth of the ordinary brood.

Bees are provided with a sting, but the drones are not so provided. The rulers are provided with stings, but they never use them; and this latter circumstance will account for the belief of some people that they have no stings at all.

Part 22

Of bees there are various species. The best kind is a little round mottled insect; another is long, and resembles the anthrena; a third is a black and flat-bellied, and is nick-named the 'robber'; a fourth kind is the drone, the largest of all, but stingless and inactive. And this proportionate size of the drone explains why some bee-masters place a net-work in front of the hives; for the network is put to keep the big drones out while it lets the little bees go in.

Of the king bees there are, as has been stated, two kinds. In every hive there are more kings than one; and a hive goes to ruin if there be too few kings, not because of anarchy thereby ensuing, but, as we are told, because these creatures contribute in some way to the generation of the common bees. A hive will go also to ruin if there be too large a number of kings in it; for the members of the hives are thereby subdivided into too many separate factions.

Whenever the spring-time is late a-coming, and when there is drought and mildew, then the progeny of the hive is small in number. But when the weather is dry they attend to the honey, and in rainy weather their attention is concentrated on the brood; and this will account for the coincidence of rich olive-harvests and abundant swarms.

The bees first work at the honeycomb, and then put the pupae in it: by the mouth, say those who hold the theory of their bringing them from elsewhere. After putting in the pupae they put in the honey for subsistence, and this they do in the summer and autumn; and, by the way, the autumn honey is the better of the two.

The honeycomb is made from flowers, and the materials for the wax they gather from the resinous gum of trees, while honey is distilled from dew, and is deposited chiefly at the risings of the constellations or when a rainbow is in the sky: and as a general rule there is no honey before the rising of the Pleiades. (The bee, then, makes the wax from flowers. The honey, however, it does not make, but merely gathers what is deposited out of the atmosphere; and as a proof of this statement we have the known fact that occasionally bee-keepers find the hives filled with honey within the space of two or three days. Furthermore, in autumn flowers are found, but honey, if it be withdrawn, is not replaced; now, after the withdrawal of the original honey, when no food or very little is in the hives, there would be a fresh stock of honey, if the bees made it from flowers.) Honey, if allowed to ripen and mature, gathers consistency; for at first it is like water and remains liquid for several days. If it be drawn off during these days it has no consistency; but it attains consistency in about twenty days. The taste of thyme-honey is discernible at once, from its peculiar sweetness and consistency.

The bee gathers from every flower that is furnished with a calyx or cup, and from all other flowers that are sweet-tasted, without doing injury to any fruit; and the juices of the flowers it takes up with the organ that resembles a tongue and carries off to the hive.

Swarms are robbed of their honey on the appearance of the wild fig. They produce the best larvae at the time the honey is a-making. The bee carries wax and bees' bread round its legs, but vomits the honey into the cell. After depositing its young, it broods over it like a bird. The grub when it is small lies slantwise in the comb, but by and by rises up straight by an effort of its own and takes food, and holds on so tightly to the honeycomb as actually to cling to it.

The young of bees and of drones is white, and from the young come the grubs; and the grubs grow into bees and drones. The egg of the king bee is reddish in colour, and its substance is about as consistent as thick honey; and from the first it is about as big as the bee that is produced from it. From the young of the king bee there is no intermediate stage, it is said, of the grub, but the bee comes at once.

Whenever the bee lays an egg in the comb there is always a drop of honey set against it. The larva of the bee gets feet and wings as soon as the cell has been stopped up with wax, and when it arrives at its completed form it breaks its membrane and flies away. It ejects excrement in the grub state, but not afterwards; that is, not until it has got out of the encasing membrane, as we have already described. If you remove the heads from off the larvae before the coming of the wings, the bees will eat them up; and if you nip off the wings from a drone and let it go, the bees will spontaneously bite off the wings from off all the remaining drones.

The bee lives for six years as a rule, as an exception for seven years. If a swarm lasts for nine years, or ten, great credit is considered due to its management.

In Pontus are found bees exceedingly white in colour, and these bees produce their honey twice a month. (The bees in Themiscyra, on the banks of the river Thermodon, build honeycombs in the ground and in hives, and these honeycombs are furnished with very little wax but with honey of great consistency; and the honeycomb, by the way, is smooth and level.) But this is not always the case with these bees, but only in the winter season; for in Pontus the ivy is abundant, and it flowers at this time of the year, and it is from the ivy-flower that they derive their honey. A white and very consistent honey is brought down from the upper country to Amisus, which is deposited by bees on trees without the employment of honeycombs: and this kind of honey is produced in other districts in Pontus.

There are bees also that construct triple honeycombs in the ground; and these honeycombs supply honey but never contain grubs. But the honeycombs in these places are not all of this sort, nor do all the bees construct them.

Part 23

Anthrenae and wasps construct combs for their young. When they have no king, but are wandering about in search of one, the anthrene constructs its comb on some high place, and the wasp inside a hole. When the anthrene and the wasp have a king, they construct their combs underground. Their combs are in all cases hexagonal like the comb of the bee. They are composed, however, not of wax, but of a bark-like filamented fibre, and the comb of the anthrene is much neater than the comb of the wasp. Like the bee, they put their young just like a drop of liquid on to the side of the cell, and the egg clings to the wall of the cell. But the eggs are not deposited in the cells simultaneously; on the contrary, in some cells are creatures big enough to fly, in others are nymphae, and in others are mere grubs. As in the case of bees, excrement is observed only in the cells where the grubs are found. As long as the creatures are in the nymph condition they are motionless, and the cell is cemented over. In the comb of the anthrene there is found in the cell of the young a drop of honey in front of it. The larvae of the anthrene and the wasp make their appearance not in the spring but in the autumn; and their growth is especially discernible in times of full moon. And, by the way, the eggs and the grubs never rest at the bottom of the cells, but always cling on to the side wall.

Part 24

There is a kind of humble-bee that builds a cone-shaped nest of clay against a stone or in some similar situation, besmearing the clay with something like spittle. And this nest or hive is exceedingly thick and hard; in point of fact, one can hardly break it open with a spike. Here the insects lay their eggs, and white grubs are produced wrapped in a black membrane. Apart from the membrane there is found some wax in the honeycomb; and this a wax is much sallower in hue than the wax in the honeycomb of the bee.

Part 25

Ants copulate and engender grubs; and these grubs attach themselves to nothing in particular, but grow on and on from small and rounded shapes until they become elongated and defined in shape: and they are engendered in spring-time.

Part 26

The land-scorpion also lays a number of egg shaped grubs, and broods over them. When the hatching is completed, the parent animal, as happens with the parent spider, is ejected and put to death by the young ones; for very often the young ones are about eleven in number.

Part 27

Spiders in all cases copulate in the way above mentioned, and generate at first small grubs. And these grubs metamorphose in their entirety, and not partially, into spiders; for, by the way, the grubs are round-shaped at the outset. And the spider, when it lays its eggs, broods over them, and in three days the eggs or grubs take definite shape.

All spiders lay their eggs in a web; but some spiders lay in a small and fine web, and others in a thick one; and some, as a rule, lay in a round-shaped case or capsule, and some are only partially enveloped in the web. The young grubs are not all developed at one and the same time into young spiders; but the moment the development takes place, the young spider makes a leap and begins to spin his web. The juice of the grub, if you squeeze it, is the same as the juice found in the spider when young; that is to say, it is thick and white.

The meadow spider lays its eggs into a web, one half of which is attached to itself and the other half is free; and on this the parent broods until the eggs are hatched. The phalangia lay their eggs in a sort of strong basket which they have woven, and brood over it until the eggs are hatched. The smooth spider is much less prolific than the phalangium or hairy spider. These phalangia, when they grow to full size, very often envelop the mother phalangium and eject and kill her; and not seldom they kill the father-phalangium as well, if they catch him: for, by the way, he has the habit of co-operating with the mother in the hatching. The brood of a single phalangium is sometimes three hundred in number. The spider attains its full growth in about four weeks.

Part 28

Grasshoppers (or locusts) copulate in the same way as other insects; that is to say, with the lesser covering the larger, for the male is smaller than the female. The females first insert the hollow tube, which they have at their tails, in the ground, and then lay their eggs: and the male, by the way, is not furnished with this tube. The females lay their eggs all in a lump together, and in one spot, so that the entire lump of eggs resembles a honeycomb. After they have laid their eggs, the eggs assume the shape of oval grubs that are enveloped by a sort of thin clay, like a membrane; in this membrane-like formation they grow on to maturity. The larva is so soft that it collapses at a touch. The larva is not placed on the surface of the ground, but a little beneath the surface; and, when it reaches maturity, it comes out of its clayey investiture in the shape of a little black grasshopper; by and by, the skin integument strips off, and it grows larger and larger.

The grasshopper lays its eggs at the close of summer, and dies after laying them. The fact is that, at the time of laying the eggs, grubs are engendered in the region of the mother grasshopper's neck; and the male grasshoppers die about the same time. In spring-time they come out of the ground; and, by the way, no grasshoppers are found in mountainous land or in poor land, but only in flat and loamy land, for the fact is they lay their eggs in cracks of the soil. During the winter their eggs remain in the ground; and with the coming of summer the last year's larva develops into the perfect grasshopper.

Part 29

The attelabi or locusts lay their eggs and die in like manner after laying them. Their eggs are subject to destruction by the autumn rains, when the rains are unusually heavy; but in seasons of drought the locusts are exceedingly numerous, from the absence of any destructive cause, since their destruction seems then to be a matter of accident and to depend on luck.

Part 30

Of the cicada there are two kinds; one, small in size, the first to come and the last to disappear; the other, large, the singing one that comes last and first disappears. Both in the small and the large species some are divided at the waist, to wit, the singing ones, and some are undivided; and these latter have no song. The large and singing cicada is by some designated the 'chirper', and the small cicada the 'tettigonium' or cicadelle. And, by the way, such of the tettigonia as are divided at the waist can sing just a little.

The cicada is not found where there are no trees; and this accounts for the fact that in the district surrounding the city of Cyrene it is not found at all in the plain country, but is found in great numbers in the neighbourhood of the city, and especially where olive-trees are growing: for an olive grove is not thickly shaded. And the cicada is not found in cold places, and consequently is not found in any grove that keeps out the sunlight.

The large and the small cicada copulate alike, belly to belly. The male discharges sperm into the female, as is the case with insects in general, and the female cicada has a cleft generative organ; and it is the female into which the male discharges the sperm.

They lay their eggs in fallow lands, boring a hole with the pointed organ they carry in the rear, as do the locusts likewise; for the locust lays its eggs in untilled lands, and this fact may account for their numbers in the territory adjacent to the city of Cyrene. The cicadae also lay their eggs in the canes on which husbandmen prop vines, perforating the canes; and also in the stalks of the squill. This brood runs into the ground. And they are most numerous in rainy weather. The grub, on attaining full size in the ground, becomes a tettigometra (or nymph), and the creature is sweetest to the taste at this stage before the husk is broken. When the summer solstice comes, the creature issues from the husk at night-time, and in a moment, as the husk breaks, the larva becomes the perfect cicada. creature, also, at once turns black in colour and harder and larger, and takes to singing. In both species, the larger and the smaller, it is the male that sings, and the female that is unvocal. At first, the males are the sweeter eating; but, after copulation, the females, as they are full then of white eggs.

If you make a sudden noise as they are flying overhead they let drop something like water. Country people, in regard to this, say that they are voiding urine, ie. that they have an excrement, and that they feed upon dew.

If you present your finger to a cicada and bend back the tip of it and then extend it again, it will endure the presentation more quietly than if you were to keep your finger outstretched altogether; and it will set to climbing your finger: for the creature is so weak-sighted that it will take to climbing your finger as though that were a moving leaf.

Part 31

Of insects that are not carnivorous but that live on the juices of living flesh, such as lice and fleas and bugs, all, without exception, generate what are called 'nits', and these nits generate nothing.

Of these insects the flea is generated out of the slightest amount of putrefying matter; for wherever there is any dry excrement, a flea is sure to be found. Bugs are generated from the moisture of living animals, as it dries up outside their bodies. Lice are generated out of the flesh of animals.

When lice are coming there is a kind of small eruption visible, unaccompanied by any discharge of purulent matter; and, if you prick an animal when in this condition at the spot of eruption, the lice jump out. In some men the appearance of lice is a disease, in cases where the body is surcharged with moisture; and, indeed, men have been known to succumb to this louse-disease, as Alcman the poet and the Syrian Pherecydes are said to have done. Moreover, in certain diseases lice appear in great abundance.

There is also a species of louse called the 'wild louse', and this is harder than the ordinary louse, and there is exceptional difficulty in getting the skin rid of it. Boys' heads are apt to be lousy, but men's in less degree; and women are more subject to lice than men. But, whenever people are troubled with lousy heads, they are less than ordinarily troubled with headache. And lice are generated in other animals than man. For birds are infested with them; and pheasants, unless they clean themselves in the dust, are actually destroyed by them. All other winged animals that are furnished with feathers are similarly infested, and all hair-coated creatures also, with the single exception of the ass, which is infested neither with lice nor with ticks.

Cattle suffer both from lice and from ticks. Sheep and goats breed ticks, but do not breed lice. Pigs breed lice large and hard. In dogs are found the flea peculiar to the animal, the Cynoroestes. In all animals that are subject to lice, the latter originate from the animals themselves. Moreover, in animals that bathe at all, lice are more than usually abundant when they change the water in which they bathe.

In the sea, lice are found on fish, but they are generated not out of the fish but out of slime; and they resemble multipedal wood-lice, only that their tail is flat. Sea-lice are uniform in shape and universal in locality, and are particularly numerous on the body of the red mullet. And all these insects are multipedal and devoid of blood.

The parasite that feeds on the tunny is found in the region of the fins; it resembles a scorpion, and is about the size of a spider. In the seas between Cyrene and Egypt there is a fish that attends on the dolphin, which is called the 'dolphin's louse'. This fish gets exceedingly fat from enjoying an abundance of food while the dolphin is out in pursuit of its prey.

Part 32

Other animalcules besides these are generated, as we have already remarked, some in wool or in articles made of wool, as the ses or clothes-moth. And these animalcules come in greater numbers if the woolen substances are dusty; and they come in especially large numbers if a spider be shut up in the cloth or wool, for the creature drinks up any moisture that may be there, and dries up the woolen substance. This grub is found also in men's clothes.

A creature is also found in wax long laid by, just as in wood, and it is the smallest of animalcules and is white in colour, and is designated the acari or mite. In books also other animalcules are found, some resembling the grubs found in garments, and some resembling tailless scorpions, but very small. As a general rule we may state that such animalcules are found in practically anything, both in dry things that are becoming moist and in moist things that are drying, provided they contain the conditions of life.

There is a grub entitled the 'faggot-bearer', as strange a creature as is known. Its head projects outside its shell, mottled in colour, and its feet are near the end or apex, as is the case with grubs in general; but the rest of its body is cased in a tunic as it were of spider's web, and there are little dry twigs about it, that look as though they had stuck by accident to the creature as it went walking about. But these twig-like formations are naturally connected with the tunic, for just as the shell is with the body of the snail so is the whole superstructure with our grub; and they do not drop off, but can only be torn off, as though they were all of a piece with him, and the removal of the tunic is as fatal to this grub as the removal of the shell would be to the snail. In course of time this grub becomes a chrysalis, as is the case with the silkworm, and lives in a motionless condition. But as yet it is not known into what winged condition it is transformed.

The fruit of the wild fig contains the psen, or fig-wasp. This creature is a grub at first; but in due time the husk peels off and the psen leaves the husk behind it and flies away, and enters into the fruit of the fig-tree through its orifice, and causes the fruit not to drop off; and with a view to this phenomenon, country folk are in the habit of tying wild figs on to fig-trees, and of planting wild fig-trees near domesticated ones.

Part 33

In the case of animals that are quadrupeds and red-blooded and oviparous, generation takes place in the spring, but copulation does not take place in a uniform season. In some cases it takes place in the spring, in others in summer time, and in others in the autumn, according as the subsequent season may be favourable for the young.

The tortoise lays eggs with a hard shell and of two colours within, like birds' eggs, and after laying them buries them in the ground and treads the ground hard over them; it then broods over the eggs on the surface of the ground, and hatches the eggs the next year. The hemys, or fresh-water tortoise, leaves the water and lays its eggs. It digs a hole of a casklike shape, and deposits therein the eggs; after rather less than thirty days it digs the eggs up again and hatches them with great rapidity, and leads its young at once off to the water. The sea-turtle lays on the ground eggs just like the eggs of domesticated birds, buries the eggs in the ground, and broods over them in the night-time. It lays a very great number of eggs, amounting at times to one hundred.

Lizards and crocodiles, terrestrial and fluvial, lay eggs on land. The eggs of lizards hatch spontaneously on land, for the lizard does not live on into the next year; in fact, the life of the animal is said not to exceed six months. The river-crocodile lays a number of eggs, sixty at the most, white in colour, and broods over them for sixty days: for, by the way, the creature is very long-lived. And the disproportion is more marked in this animal than in any other between the smallness of the original egg and the huge size of the full-grown animal. For the egg is not larger than that of the goose, and the young crocodile is small, answering to the egg in size, but the full-grown animal attains the length of twenty-six feet; in fact, it is actually stated that the animal goes on growing to the end of its days.

Part 34

With regard to serpents or snakes, the viper is externally viviparous, having been previously oviparous internally. The egg, as with the egg of fishes, is uniform in colour and soft-skinned. The young serpent grows on the surface of the egg, and, like the young of fishes, has no shell-like envelopment. The young of the viper is born inside a membrane that bursts from off the young creature in three days; and at times the young viper eats its way out from the inside of the egg. The mother viper brings forth all its young in one day, twenty in number, and one at a time. The other serpents are externally oviparous, and their eggs are strung on to one another like a lady's necklace; after the dam has laid her eggs in the ground she broods over them, and hatches the eggs in the following year.

Book VI

Part 1

SO much for the generative processes in snakes and insects, and also in oviparous quadrupeds. Birds without exception lay eggs, but the pairing season and the times of parturition are not alike for all. Some birds couple and lay at almost any time in the year, as for instance the barn-door hen and the pigeon: the former of these coupling and laying during the entire year, with the exception of the month before and the month after the winter solstice. Some hens, even in the high breeds, lay a large quantity of eggs before brooding, amounting to as many as sixty; and, by the way, the higher breeds are less prolific than the inferior ones. The Adrian hens are small-sized, but they lay every day; they are cross-tempered, and often kill their chickens; they are of all colours. Some domesticated hens lay twice a day; indeed, instances have been known where hens, after exhibiting extreme fecundity, have died suddenly. Hens, then, lay eggs, as has been stated, at all times indiscriminately; the pigeon, the ring-dove, the turtle-dove, and the stock-dove lay twice a year, and the pigeon actually lays ten times a year. The great majority of birds lay during the spring-time. Some birds are prolific, and prolific in either of two ways-either by laying often, as the pigeon, or by laying many eggs at a sitting, as the barn-door hen. All birds of prey, or birds with crooked talons, are unprolific, except the kestrel: this bird is the most prolific of birds of prey; as many as four eggs have been observed in the nest, and occasionally it lays even more.

Birds in general lay their eggs in nests, but such as are disqualified for flight, as the partridge and the quail, do not lay them in nests but on the ground, and cover them over with loose material. The same is the case with the lark and the tetrix. These birds hatch in sheltered places; but the bird called merops in Boeotia, alone of all birds, burrows into holes in the ground and hatches there.

Thrushes, like swallows, build nests of clay, on high trees, and build them in rows all close together, so that from their continuity the structure resembles a necklace of nests. Of all birds that hatch for themselves the hoopoe is the only one that builds no nest whatever; it gets into the hollow of the trunk of a tree, and lays its eggs there without making any sort of nest. The circus builds either under a dwelling-roof or on cliffs. The tetrix, called ourax in Athens, builds neither on the ground nor on trees, but on low-lying shrubs.

Part 2

The egg in the case of all birds alike is hard-shelled, if it be the produce of copulation and be laid by a healthy hen-for some hens lay soft eggs. The interior of the egg is of two colours, and the white part is outside and the yellow part within.

The eggs of birds that frequent rivers and marshes differ from those of birds that live on dry land; that is to say, the eggs of waterbirds have comparatively more of the yellow or yolk and less of the white. Eggs vary in colour according to their kind. Some eggs are white, as those of the pigeon and of the partridge; others are yellowish, as the eggs of marsh birds; in some cases the eggs are mottled, as the eggs of the guinea-fowl and the pheasant; while the eggs of the kestrel are red, like vermilion.

Eggs are not symmetrically shaped at both ends: in other words, one end is comparatively sharp, and the other end is comparatively blunt; and it is the latter end that protrudes first at the time of laying. Long and pointed eggs are female; those that are round, or more rounded at the narrow end, are male. Eggs are hatched by the incubation of the mother-bird. In some cases, as in Egypt, they are hatched spontaneously in the ground, by being buried in dung heaps. A story is told of a toper in Syracuse, how he used to put eggs into the ground under his rush-mat and to keep on drinking until he hatched them. Instances have occurred of eggs being deposited in warm vessels and getting hatched spontaneously.

The sperm of birds, as of animals in general, is white. After the female has submitted to the male, she draws up the sperm to underneath her midriff. At first it is little in size and white in colour; by and by it is red, the colour of blood; as it grows, it becomes pale and yellow all over. When at length it is getting ripe for hatching, it is subject to differentiation of substance, and the yolk gathers together within and the white settles round it on the outside. When the full time is come, the egg detaches itself and protrudes, changing from soft to hard with such temporal exactitude that, whereas it is not hard during the process of protrusion, it hardens immediately after the process is completed: that is if there be no concomitant pathological circumstances. Cases have occurred where substances resembling the egg at a critical point of its growth-that is, when it is yellow all over, as the yolk is subsequently-have been found in the cock when cut open, underneath his midriff, just where the hen has her eggs; and these are entirely yellow in appearance and of the same size as ordinary eggs. Such phenomena are regarded as unnatural and portentous.

Such as affirm that wind-eggs are the residua of eggs previously begotten from copulation are mistaken in this assertion, for we have cases well authenticated where chickens of the common hen and goose have laid wind-eggs without ever having been subjected to copulation. Wind-eggs are smaller, less palatable, and more liquid than true eggs, and are produced in greater numbers. When they are put under the mother bird, the liquid contents never coagulate, but both the yellow and the white remain as they were. Wind-eggs are laid by a number of birds: as for instance by the common hen, the hen partridge, the hen pigeon, the peahen, the goose, and the vulpanser. Eggs are hatched under brooding hens more rapidly in summer than in winter; that is to say, hens hatch in eighteen days in summer, but occasionally in winter take as many as twenty-five. And by the way for brooding purposes some birds make better mothers than others. If it thunders while a hen-bird is brooding, the eggs get addled. Wind-eggs that are called by some cynosura and uria are produced chiefly in summer. Wind-eggs are called by some zephyr-eggs, because at spring-time hen-birds are observed to inhale the breezes; they do the same if they be stroked in a peculiar way by hand. Wind-eggs can turn into fertile eggs, and eggs due to previous copulation can change breed, if before the change of the yellow to the white the hen that contains wind-eggs, or eggs begotten of copulation be trodden by another cock-bird. Under these circumstances the wind-eggs turn into fertile eggs, and the previously impregnated eggs follow the breed of the impregnator; but if the latter impregnation takes place during the change of the yellow to the white, then no change in the egg takes place: the wind-egg does not become a true egg, and the true egg does not take on the breed of the latter impregnator. If when the egg-substance is small copulation be intermitted, the previously existing egg-substance exhibits no increase; but if the hen be again submitted to the male the increase in size proceeds with rapidity.

The yolk and the white are diverse not only in colour but also in properties. Thus, the yolk congeals under the influence of cold, whereas the white instead of congealing is inclined rather to liquefy. Again, the white stiffens under the influence of fire, whereas the yolk does not stiffen; but, unless it be burnt through and through, it remains soft, and in point of fact is inclined to set or to harden more from the boiling than from the roasting of the egg. The yolk and the white are separated by a membrane from one another. The so-called 'hail-stones', or treadles, that are found at the extremity of the yellow in no way contribute towards generation, as some erroneously suppose: they are two in number, one below and the other above. If you take out of the shells a number of yolks and a number of whites and pour them into a sauce pan and boil them slowly over a low fire, the yolks will gather into the centre and the whites will set all around them.

Young hens are the first to lay, and they do so at the beginning of spring and lay more eggs than the older hens, but the eggs of the younger hens are comparatively small. As a general rule, if hens get no brooding they pine and sicken. After copulation hens shiver and shake themselves, and often kick rubbish about all round them-and this, by the way, they do sometimes after laying-whereas pigeons trail their rumps on the ground, and geese dive under the water. Conception of the true egg and conformation of the wind-egg take place rapidly with most birds; as for instance with the hen-partridge when in heat. The fact is that, when she stands to windward and within scent of the male, she conceives, and becomes useless for decoy purposes: for, by the way, the partridge appears to have a very acute sense of smell.

The generation of the egg after copulation and the generation of the chick from the subsequent hatching of the egg are not brought about within equal periods for all birds, but differ as to time according to the size of the parent-birds. The egg of the common hen after copulation sets and matures in ten days a general rule; the egg of the pigeon in a somewhat lesser period. Pigeons have the faculty of holding back the egg at the very moment of parturition; if a hen pigeon be put about by any one, for instance if it be disturbed on its nest, or have a feather plucked out, or sustain any other annoyance or disturbance, then even though she had made up her mind to lay she can keep the egg back in abeyance. A singular phenomenon is observed in pigeons with regard to pairing: that is, they kiss one another just when the male is on the point of mounting the female, and without this preliminary the male would decline to perform his function. With the older males the preliminary kiss is only given to begin with, and subsequently he mounts without previously kissing; with younger males the preliminary is never omitted. Another singularity in these birds is that the hens tread one another when a cock is not forthcoming, after kissing one another just as takes place in the normal pairing. Though they do not impregnate one another they lay more eggs under these than under ordinary circumstances; no chicks, however, result therefrom, but all such eggs are wind-eggs.

Part 3

Generation from the egg proceeds in an identical manner with all birds, but the full periods from conception to birth differ, as has been said. With the common hen after three days and three nights there is the first indication of the embryo; with larger birds the interval being longer, with smaller birds shorter. Meanwhile the yolk comes into being, rising towards the sharp end, where the primal element of the egg is situated, and where the egg gets hatched; and the heart appears, like a speck of blood, in the white of the egg. This point beats and moves as though endowed with life, and from it two vein-ducts with blood in them trend in a convoluted course (as the egg substance goes on growing, towards each of the two circumjacent integuments); and a membrane carrying bloody fibres now envelops the yolk, leading off from the vein-ducts. A little afterwards the body is differentiated, at first very small and white. The head is clearly distinguished, and in it the eyes, swollen out to a great extent. This condition of the eyes lat on for a good while, as it is only by degrees that they diminish in size and collapse. At the outset the under portion of the body appears insignificant in comparison with the upper portion. Of the two ducts that lead from the heart, the one proceeds towards the circumjacent integument, and the other, like a navel-string, towards the yolk. The life-element of the chick is in the white of the egg, and the nutriment comes through the navel-string out of the yolk.

When the egg is now ten days old the chick and all its parts are distinctly visible. The head is still larger than the rest of its body, and the eyes larger than the head, but still devoid of vision. The eyes, if removed about this time, are found to be larger than beans, and black; if the cuticle be peeled off them there is a white and cold liquid inside, quite glittering in the sunlight, but there is no hard substance whatsoever. Such is the condition of the head and eyes. At this time also the larger internal organs are visible, as also the stomach and the arrangement of the viscera; and veins that seem to proceed from the heart are now close to the navel. From the navel there stretch a pair of veins; one towards the membrane that envelops the yolk (and, by the way, the yolk is now liquid, or more so than is normal), and the other towards that membrane which envelops collectively the membrane wherein the chick lies, the membrane of the yolk, and the intervening liquid. (For, as the chick grows, little by little one part of the yolk goes upward, and another part downward, and the white liquid is between them; and the white of the egg is underneath the lower part of the yolk, as it was at the outset.) On the tenth day the white is at the extreme outer surface, reduced in amount, glutinous, firm in substance, and sallow in colour.

The disposition of the several constituent parts is as follows. First and outermost comes the membrane of the egg, not that of the shell, but underneath it. Inside this membrane is a white liquid; then comes the chick, and a membrane round about it, separating it off so as to keep the chick free from the liquid; next after the chick comes the yolk, into which one of the two veins was described as leading, the other one leading into the enveloping white substance. (A membrane with a liquid resembling serum envelops the entire structure. Then comes another membrane right round the embryo, as has been described, separating it off against the liquid. Underneath this comes the yolk, enveloped in another membrane (into which yolk proceeds the navel-string that leads from the heart and the big vein), so as to keep the embryo free of both liquids.)

About the twentieth day, if you open the egg and touch the chick, it moves inside and chirps; and it is already coming to be covered with down, when, after the twentieth day is ast, the chick begins to break the shell. The head is situated over the right leg close to the flank, and the wing is placed over the head; and about this time is plain to be seen the membrane resembling an after-birth that comes next after the outermost membrane of the shell, into which membrane the one of the navel-strings was described as leading (and, by the way, the chick in its entirety is now within it), and so also is the other membrane resembling an after-birth, namely that surrounding the yolk, into which the second navel-string was described as leading; and both of them were described as being connected with the heart and the big vein. At this conjuncture the navel-string that leads to the outer afterbirth collapses and becomes detached from the chick, and the membrane that leads into the yolk is fastened on to the thin gut of the creature, and by this time a considerable amount of the yolk is inside the chick and a yellow sediment is in its stomach. About this time it discharges residuum in the direction of the outer after-birth, and has residuum inside its stomach; and the outer residuum is white (and there comes a white substance inside). By and by the yolk, diminishing gradually in size, at length becomes entirely used up and comprehended within the chick (so that, ten days after hatching, if you cut open the chick, a small remnant of the yolk is still left in connexion with the gut), but it is detached from the navel, and there is nothing in the interval between, but it has been used up entirely. During the period above referred to the chick sleeps, wakes up, makes a move and looks up and Chirps; and the heart and the navel together palpitate as though the creature were respiring. So much as to generation from the egg in the case of birds.

Birds lay some eggs that are unfruitful, even eggs that are the result of copulation, and no life comes from such eggs by incubation; and this phenomenon is observed especially with pigeons.

Twin eggs have two yolks. In some twin eggs a thin partition of white intervenes to prevent the yolks mixing with each other, but some twin eggs are unprovided with such partition, and the yokes run into one another. There are some hens that lay nothing but twin eggs, and in their case the phenomenon regarding the yolks has been observed. For instance, a hen has been known to lay eighteen eggs, and to hatch twins out of them all, except those that were wind-eggs; the rest were fertile (though, by the way, one of the twins is always bigger than the other), but the eighteenth was abnormal or monstrous.

Part 4

Birds of the pigeon kind, such as the ringdove and the turtle-dove, lay two eggs at a time; that is to say, they do so as a general rule, and they never lay more than three. The pigeon, as has been said, lays at all seasons; the ring-dove and the turtle-dove lay in the springtime, and they never lay more than twice in the same season. The hen-bird lays the second pair of eggs when the first pair happens to have been destroyed, for many of the hen-pigeons destroy the first brood. The hen-pigeon, as has been said, occasionally lays three eggs, but it never rears more than two chicks, and sometimes rears only one; and the odd one is always a wind-egg.

Very few birds propagate within their first year. All birds, after once they have begun laying, keep on having eggs, though in the case of some birds it is difficult to detect the fact from the minute size of the creature.

The pigeon, as a rule, lays a male and a female egg, and generally lays the male egg first; after laying it allows a day's interval to ensue and then lays the second egg. The male takes its turn of sitting during the daytime; the female sits during the night. The first-laid egg is hatched and brought to birth within twenty days; and the mother bird pecks a hole in the egg the day before she hatches it out. The two parent birds brood for some time over the chicks in the way in which they brooded previously over the eggs. In all connected with the rearing of the young the female parent is more cross-tempered than the male, as is the case with most animals after parturition. The hens lay as many as ten times in the year; occasional instances have been known of their laying eleven times, and in Egypt they actually lay twelve times. The pigeon, male and female, couples within the year; in fact, it couples when only six months old. Some assert that ringdoves and turtle-doves pair and procreate when only three months old, and instance their superabundant numbers by way of proof of the assertion. The hen-pigeon carries her eggs fourteen days; for as many more days the parent birds hatch the eggs; by the end of another fourteen days the chicks are so far capable of flight as to be overtaken with difficulty. (The ring-dove, according to all accounts, lives up to forty years. The partridge lives over sixteen.) (After one brood the pigeon is ready for another within thirty days.)

Part 5

The vulture builds its nest on inaccessible cliffs; for which reason its nest and young are rarely seen. And therefore Herodorus, father of Bryson the Sophist, declares that vultures belong to some foreign country unknown to us, stating as a proof of the assertion that no one has ever seen a vulture's nest, and also that vultures in great numbers make a sudden appearance in the rear of armies. However, difficult as it is to get a sight of it, a vulture's nest has been seen. The vulture lays two eggs.

(Carnivorous birds in general are observed to lay but once a year. The swallow is the only carnivorous bird that builds a nest twice. If you prick out the eyes of swallow chicks while they are yet young, the birds will get well again and will see by and by.)

Part 6

The eagle lays three eggs and hatches two of them, as it is said in the verses ascribed to Musaeus:

That lays three, hatches two, and cares for one.

This is the case in most instances, though occasionally a brood of three has been observed. As the young ones grow, the mother becomes wearied with feeding them and extrudes one of the pair from the nest. At the same time the bird is said to abstain from food, to avoid harrying the young of wild animals. That is to say, its wings blanch, and for some days its talons get turned awry. It is in consequence about this time cross-tempered to its own young. The phene is said to rear the young one that has been expelled the nest. The eagle broods for about thirty days.

The hatching period is about the same for the larger birds, such as the goose and the great bustard; for the middle-sized birds it extends over about twenty days, as in the case of the kite and the hawk. The kite in general lays two eggs, but occasionally rears three young ones. The so-called aegolius at times rears four. It is not true that, as some aver, the raven lays only two eggs; it lays a larger number. It broods for about twenty days and then extrudes its young. Other birds perform the same operation; at all events mother birds that lay several eggs often extrude one of their young.

Birds of the eagle species are not alike in the treatment of their young. The white-tailed eagle is cross, the black eagle is affectionate in the feeding of the young; though, by the way, all birds of prey, when their brood is rather forward in being able to fly, beat and extrude them from the nest.

The majority of birds other than birds of prey, as has been said, also act in this manner, and after feeding their young take no further care of them; but the crow is an exception. This bird for a considerable time takes charge of her young; for, even when her young can fly, she flies alongside of them and supplies them with food.

Part 7

The cuckoo is said by some to be a hawk transformed, because at the time of the cuckoo's coming, the hawk, which it resembles, is never seen; and indeed it is only for a few days that you will see hawks about when the cuckoo's note sounds early in the season. The cuckoo appears only for a short time in summer, and in winter disappears. The hawk has crooked talons, which the cuckoo has not; neither with regard to the head does the cuckoo resemble the hawk. In point of fact, both as regards the head and the claws it more resembles the pigeon. However, in colour and in colour alone it does resemble the hawk, only that the markings of the hawk are striped, and of the cuckoo mottled. And, by the way, in size and flight it resembles the smallest of the hawk tribe, which bird disappears as a rule about the time of the appearance of the cuckoo, though the two have been seen simultaneously. The cuckoo has been seen to be preyed on by the hawk; and this never happens between birds of the same species. They say no one has ever seen the young of the cuckoo. The bird eggs, but does not build a nest. Sometimes it lays its eggs in the nest of a smaller bird after first devouring the eggs of this bird; it lays by preference in the nest of the ringdove, after first devouring the eggs of the pigeon. (It occasionally lays two, but usually one.) It lays also in the nest of the hypolais, and the hypolais hatches and rears the brood. It is about this time that the bird becomes fat and palatable. (The young of hawks also get palatable and fat. One species builds a nest in the wilderness and on sheer and inaccessible cliffs.)

Part 8

With most birds, as has been said of the pigeon, the hatching is carried on by the male and the female in turns: with some birds, however, the male only sits long enough to allow the female to provide herself with food. In the goose tribe the female alone incubates, and after once sitting on the eggs she continues brooding until they are hatched.

The nests of all marsh-birds are built in districts fenny and well supplied with grass; consequently, the mother-bird while sitting quiet on her eggs can provide herself with food without having to submit to absolute fasting.

With the crow also the female alone broods, and broods throughout the whole period; the male bird supports the female, bringing her food and feeding her. The female of the ring-dove begins to brood in the afternoon and broods through the entire night until breakfast-time of the following day; the male broods during the rest of the time. Partridges build a nest in two compartments; the male broods on the one and the female on the other. After hatching, each of the parent birds rears its brood. But the male, when he first takes his young out of the nest, treads them.

Part 9

Peafowl live for about twenty-five years, breed about the third year, and at the same time take on their spangled plumage. They hatch their eggs within thirty days or rather more. The peahen lays but once a year, and lays twelve eggs, or may be a slightly lesser number: she does not lay all the eggs there and then one after the other, but at intervals of two or three days. Such as lay for the first time lay about eight eggs. The peahen lays wind-eggs. They pair in the spring; and laying begins immediately after pairing. The bird moults when the earliest trees are shedding their leaves, and recovers its plumage when the same trees are recovering their foliage. People that rear peafowl put the eggs under the barn-door hen, owing to the fact that when the peahen is brooding over them the peacock attacks her and tries to trample on them; owing to this circumstance some birds of wild varieties run away from the males and lay their eggs and brood in solitude. Only two eggs are put under a barn-door hen, for she could not brood over and hatch a large number. They take every precaution, by supplying her with food, to prevent her going off the eggs and discontinuing the brooding.

With male birds about pairing time the testicles are obviously larger than at other times, and this is conspicuously the case with the more salacious birds, such as the barn-door cock and the cock partridge; the peculiarity is less conspicuous in such birds as are intermittent in regard to pairing.

Part 10

So much for the conception and generation of birds. It has been previously stated that fishes are not all oviparous. Fishes of the cartilaginous genus are viviparous; the rest are oviparous. And cartilaginous fishes are first oviparous internally and subsequently viviparous; they rear the embryos internally, the batrachus or fishing-frog being an exception.

Fishes also, as was above stated, are provided with wombs, and wombs of diverse kinds. The oviparous genera have wombs bifurcate in shape and low down in position; the cartilaginous genus have wombs shaped like those of O birds. The womb, however, in the cartilaginous fishes differs in this respect from the womb of birds, that with some cartilaginous fishes the eggs do not settle close to the diaphragm but middle-ways along the backbone, and as they grow they shift their position.

The egg with all fishes is not of two colours within but is of even hue; and the colour is nearer to white than to yellow, and that both when the young is inside it and previously as well.

Development from the egg in fishes differs from that in birds in this respect, that it does not exhibit that one of the two navel-strings that leads off to the membrane that lies close under the shell, while it does exhibit that one of the two that in the case of birds leads off to the yolk. In a general way the rest of the development from the egg onwards is identical in birds and fishes. That is to say, development takes place at the upper part of the egg, and the veins extend in like manner, at first from the heart; and at first the head, the eyes, and the upper parts are largest; and as the creature grows the egg-substance decreases and eventually disappears, and becomes absorbed within the embryo, just as takes place with the yolk in birds.

The navel-string is attached a little way below the aperture of the belly. When the creatures are young the navel-string is long, but as they grow it diminishes in size; at length it gets small and becomes incorporated, as was described in the case of birds. The embryo and the egg are enveloped by a common membrane, and just under this is another membrane that envelops the embryo by itself; and in between the two membranes is a liquid. The food inside the stomach of the little fishes resembles that inside the stomach of young chicks, and is partly white and partly yellow.

As regards the shape of the womb, the reader is referred to my treatise on Anatomy. The womb, however, is diverse in diverse fishes, as for instance in the sharks as compared one with another or as compared with the skate. That is to say, in some sharks the eggs adhere in the middle of the womb round about the backbone, as has been stated, and this is the case with the dog-fish; as the eggs grow they shift their place; and since the womb is bifurcate and adheres to the midriff, as in the rest of similar creatures, the eggs pass into one or other of the two compartments. This womb and the womb of the other sharks exhibit, as you go a little way off from the midriff, something resembling white breasts, which never make their appearance unless there be conception.

Dog-fish and skate have a kind of egg-shell, in which is found an egg-like liquid. The shape of the egg-shell resembles the tongue of a bagpipe, and hair-like ducts are attached to the shell. With the dog-fish which is called by some the 'dappled shark', the young are born when the shell-formation breaks in pieces and falls out; with the ray, after it has laid the egg the shell-formation breaks up and the young move out. The spiny dog-fish has its close to the midriff above the breast like formations; when the egg descends, as soon as it gets detached the young is born. The mode of generation is the same in the case of the fox-shark.

The so-called smooth shark has its eggs in betwixt the wombs like the dog-fish; these eggs shift into each of the two horns of the womb and descend, and the young develop with the navel-string attached to the womb, so that, as the egg-substance gets used up, the embryo is sustained to all appearance just as in the case of quadrupeds. The navel-string is long and adheres to the under part of the womb (each navel-string being attached as it were by a sucker), and also to the centre of the embryo in the place where the liver is situated. If the embryo be cut open, even though it has the egg-substance no longer, the food inside is egg-like in appearance. Each embryo, as in the case of quadrupeds, is provided with a chorion and separate membranes. When young the embryo has its head upwards, but downwards when it gets strong and is completed in form. Males are generated on the left-hand side of the womb, and females on the right-hand side, and males and females on the same side together. If the embryo be cut open, then, as with quadrupeds, such internal organs as it is furnished with, as for instance the liver, are found to be large and supplied with blood.

All cartilaginous fishes have at one and the same time eggs above close to the midriff (some larger, some smaller), in considerable numbers, and also embryos lower down. And this circumstance leads many to suppose that fishes of this species pair and bear young every month, inasmuch as they do not produce all their young at once, but now and again and over a lengthened period. But such eggs as have come down below within the womb are simultaneously ripened and completed in growth.

Dog-fish in general can extrude and take in again their young, as can also the angel-fish and the electric ray-and, by the way, a large electric ray has been seen with about eighty embryos inside it-but the spiny dogfish is an exception to the rule, being prevented by the spine of the young fish from so doing. Of the flat cartilaginous fish, the trygon and the ray cannot extrude and take in again in consequence of the roughness of the tails of the young. The batrachus or fishing-frog also is unable to take in its young owing to the size of the head and the prickles; and, by the way, as was previously remarked, it is the only one of these fishes that is not viviparous.

So much for the varieties of the cartilaginous species and for their modes of generation from the egg.

Part 11

At the breeding season the sperm-ducts of the male are filled with sperm, so much so that if they be squeezed the sperm flows out spontaneously as a white fluid; the ducts are bifurcate, and start from the midriff and the great vein. About this period the sperm-ducts of the male are quite distinct (from the womb of the female) but at any other than the actual breeding time their distinctness is not obvious to a non-expert. The fact is that in certain fishes at certain times these organs are imperceptible, as was stated regarding the testicles of birds.

Among other distinctions observed between the thoric ducts and the womb-ducts is the circumstance that the thoric ducts are attached to the loins, while the womb-ducts move about freely and are attached by a thin membrane. The particulars regarding the thoric ducts may be studied by a reference to the diagrams in my treatise on Anatomy.

Cartilaginous fishes are capable of superfetation, and their period of gestation is six months at the longest. The so-called starry dogfish bears young the most frequently; in other words it bears twice a month. The breeding season is in the month of Maemacterion. The dog-fish as a general rule bear twice in the year, with the exception of the little dog-fish, which bears only once a year. Some of them bring forth in the springtime. The rhine, or angel-fish, bears its first brood in the springtime, and its second in the autumn, about the winter setting of the Pleiades; the second brood is the stronger of the two. The electric ray brings forth in the late autumn.

Cartilaginous fishes come out from the main seas and deep waters towards the shore and there bring forth their young, and they do so for the sake of warmth and by way of protection for their young.

Observations would lead to the general rule that no one variety of fish pairs with another variety. The angel-fish, however, and the batus or skate appear to pair with one another; for there is a fish called the rhinobatus, with the head and front parts of the skate and the after parts of the rhine or angel-fish, just as though it were made up of both fishes together.

Sharks then and their congeners, as the fox-shark and the dog-fish, and the flat fishes, such as the electric ray, the ray, the smooth skate, and the trygon, are first oviparous and then viviparous in the way above mentioned, (as are also the saw-fish and the ox-ray.)

Part 12

The dolphin, the whale, and all the rest of the Cetacea, all, that is to say, that are provided with a blow-hole instead of gills, are viviparous. That is to say, no one of all these fishes is ever seen to be supplied with eggs, but directly with an embryo from whose differentiation comes the fish, just as in the case of mankind and the viviparous quadrupeds.

The dolphin bears one at a time generally, but occasionally two. The whale bears one or at the most two, generally two. The porpoise in this respect resembles the dolphin, and, by the way, it is in form like a little dolphin, and is found in the Euxine; it differs, however, from the dolphin as being less in size and broader in the back; its colour is leaden-black. Many people are of opinion that the porpoise is a variety of the dolphin.

All creatures that have a blow-hole respire and inspire, for they are provided with lungs. The dolphin has been seen asleep with his nose above water, and when asleep he snores.

The dolphin and the porpoise are provided with milk, and suckle their young. They also take their young, when small, inside them. The young of the dolphin grow rapidly, being full grown at ten years of age. Its period of gestation is ten months. It brings forth its young summer, and never at any other season; (and, singularly enough, under the Dogstar it disappears for about thirty days). Its young accompany it for a considerable period; and, in fact, the creature is remarkable for the strength of its parental affection. It lives for many years; some are known to have lived for more than twenty-five, and some for thirty years; the fact is fishermen nick their tails sometimes and set them adrift again, and by this expedient their ages are ascertained.

The seal is an amphibious animal: that is to say, it cannot take in water, but breathes and sleeps and brings forth on dry land-only close to the shore-as being an animal furnished with feet; it spends, however, the greater part of its time in the sea and derives its food from it, so that it must be classed in the category of marine animals. It is viviparous by immediate conception and brings forth its young alive, and exhibits an after-birth and all else just like a ewe. It bears one or two at a time, and three at the most. It has two teats, and suckles its young like a quadruped. Like the human species it brings forth at all seasons of the year, but especially at the time when the earliest kids are forthcoming. It conducts its young ones, when they are about twelve days old, over and over again during the day down to the sea, accustoming them by slow degrees to the water. It slips down steep places instead of walking, from the fact that it cannot steady itself by its feet. It can contract and draw itself in, for it is fleshy and soft and its bones are gristly. Owing to the flabbiness of its body it is difficult to kill a seal by a blow, unless you strike it on the temple. It looks like a cow. The female in regard to its genital organs resembles the female of the ray; in all other respects it resembles the female of the human species.

So much for the phenomena of generation and of parturition in animals that live in water and are viviparous either internally or externally.

Part 13

Oviparous fishes have their womb bifurcate and placed low down, as was said previously-and, by the way, all scaly fish are oviparous, as the bass, the mullet, the grey mullet, and the etelis, and all the so-called white-fish, and all the smooth or slippery fish except the eel-and their roe is of a crumbling or granular substance. This appearance is due to the fact that the whole womb of such fishes is full of eggs, so that in little fishes there seem to be only a couple of eggs there; for in small fishes the womb is indistinguishable, from its diminutive size and thin contexture. The pairing of fishes has been discussed previously.

Fish for the most part are divided into males and females, but one is puzzled to account for the erythrinus and the channa, for specimens of these species are never caught except in a condition of pregnancy.

With such fish as pair, eggs are the result of copulation, but such fish have them also without copulation; and this is shown in the case of some river-fish, for the minnow has eggs when quite small,-almost, one may say, as soon as it is born. These fishes shed their eggs little by little, and, as is stated, the males swallow the greater part of them, and some portion of them goes to waste in the water; but such of the eggs as the female deposits on the spawning beds are saved. If all the eggs were preserved, each species would be infinite in number. The greater number of these eggs so deposited are not productive, but only those over which the male sheds the milt or sperm; for when the female has laid her eggs, the male follows and sheds its sperm over them, and from all the eggs so besprinkled young fishes proceed, while the rest are left to their fate.

The same phenomenon is observed in the case of molluscs also; for in the case of the cuttlefish or sepia, after the female has deposited her eggs, the male besprinkles them. It is highly probable that a similar phenomenon takes place in regard to molluscs in general, though up to the present time the phenomenon has been observed only in the case of the cuttlefish.

Fish deposit their eggs close in to shore, the goby close to stones; and, by the way, the spawn of the goby is flat and crumbly. Fish in general so deposit their eggs; for the water close in to shore is warm and is better supplied with food than the outer sea, and serves as a protection to the spawn against the voracity of the larger fish. And it is for this reason that in the Euxine most fishes spawn near the mouth of the river Thermodon, because the locality is sheltered, genial, and supplied with fresh water.

Oviparous fish as a rule spawn only once a year. The little phycis or black goby is an exception, as it spawns twice; the male of the black goby differs from the female as being blacker and having larger scales.

Fish then in general produce their young by copulation, and lay their eggs; but the pipefish, as some call it, when the time of parturition arrives, bursts in two, and the eggs escape out. For the fish has a diaphysis or cloven growth under the belly and abdomen (like the blind snakes), and, after it has spawned by the splitting of this diaphysis, the sides of the split grow together again.

Development from the egg takes place similarly with fishes that are oviparous internally and with fishes that are oviparous externally; that is to say, the embryo comes at the upper end of the egg and is enveloped in a membrane, and the eyes, large and spherical, are the first organs visible. From this circumstance it is plain that the assertion is untenable which is made by some writers, to wit, that the young of oviparous fish are generated like the grubs of worms; for the opposite phenomena are observed in the case of these grubs, in that their lower extremities are the larger at the outset, and that the eyes and the head appear later on. After the egg has been used up, the young fish are like tadpoles in shape, and at first, without taking any nutriment, they grow by sustenance derived from the juice oozing from the egg; by and by, they are nourished up to full growth by the river-waters.

When the Euxine is 'purged' a substance called phycus is carried into the Hellespont, and this substance is of a pale yellow colour. Some writers aver that it is the flower of the phycus, from which rouge is made; it comes at the beginning of summer. Oysters and the small fish of these localities feed on this substance, and some of the inhabitants of these maritime districts say that the purple murex derives its peculiar colour from it.

Part 14

Marsh-fishes and river-fishes conceive at the age of five months as a general rule, and deposit their spawn towards the close of the year without exception. And with these fish, like as with the marine fish, the female does not void all her eggs at one time, nor the male his sperm; but they are at all times more or less provided, the female with eggs, and the male with sperm. The-carp spawns as the seasons come round, five or six times, and follows in spawning the rising of the greater constellations. The chalcis spawns three times, and the other fish once only in the year. They all spawn in pools left by the overflowing of rivers, and near to reedy places in marshes; as for instance the phoxinus or minnow and the perch.

The glanis or sheat-fish and the perch deposit their spawn in one continuous string, like the frog; so continuous, in fact, is the convoluted spawn of the perch that, by reason of its smoothness, the fishermen in the marshes can unwind it off the reeds like threads off a reel. The larger individuals of the sheat-fish spawn in deep waters, some in water of a fathom's depth, the smaller in shallower water, generally close to the roots of the willow or of some other tree, or close to reeds or to moss. At times these fishes intertwine with one another, a big with a little one, and bring into juxtaposition the ducts-which some writers designate as navels-at the point where they emit the generative products and discharge the egg in the case of the female and the milt in the case of the male. Such eggs as are besprinkled with the milt grow, in a day or thereabouts, whiter and larger, and in a little while afterwards the fish's eyes become visible for these organs in all fishes, as for that matter in all other animals, are early conspicuous and seem disproportionately big. But such eggs as the milt fails to touch remain, as with marine fishes, useless and infertile. From the fertile eggs, as the little fish grow, a kind of sheath detaches itself; this is a membrane that envelops the egg and the young fish. When the milt has mingled with the eggs, the resulting product becomes very sticky or viscous, and adheres to the roots of trees or wherever it may have been laid. The male keeps on guard at the principal spawning-place, and the female after spawning goes away.

In the case of the sheat-fish the growth from the egg is exceptionally slow, and, in consequence, the male has to keep watch for forty or fifty days to prevent the-spawn being devoured by such little fish as chance to come by. Next in point of slowness is the generation of the carp. As with fish in general, so even with these, the spawn thus protected disappears and gets lost rapidly.

In the case of some of the smaller fish when they are only three days old young fish are generated. Eggs touched by the male sperm take on increase both the same day and also later. The egg of the sheat-fish is as big as a vetch-seed; the egg of the carp and of the carp-species as big as a millet-seed.

These fish then spawn and generate in the way here described. The chalcis, however, spawns in deep water in dense shoals of fish; and the so-called tilon spawns near to beaches in sheltered spots in shoals likewise. The carp, the baleros, and fishes in general push eagerly into the shallows for the purpose of spawning, and very often thirteen or fourteen males are seen following a single female. When the female deposits her spawn and departs, the males follow on and shed the milt. The greater portion of the spawn gets wasted; because, owing to the fact that the female moves about while spawning, the spawn scatters, or so much of it as is caught in the stream and does not get entangled with some rubbish. For, with the exception of the sheatfish, no fish keeps on guard; unless, by the way, it be the carp, which is said to remain on guard, if it so happen that its spawn lies in a solid mass.

All male fish are supplied with milt, excepting the eel: with the eel, the male is devoid of milt, and the female of spawn. The mullet goes up from the sea to marshes and rivers; the eels, on the contrary, make their way down from the marshes and rivers to the sea.

Part 15

The great majority of fish, then, as has been stated, proceed from eggs. However, there are some fish that proceed from mud and sand, even of those kinds that proceed also from pairing and the egg. This occurs in ponds here and there, and especially in a pond in the neighbourhood of Cnidos. This pond, it is said, at one time ran dry about the rising of the Dogstar, and the mud had all dried up; at the first fall of the rains there was a show of water in the pond, and on the first appearance of the water shoals of tiny fish were found in the pond. The fish in question was a kind of mullet, one which does not proceed from normal pairing, about the size of a small sprat, and not one of these fishes was provided with either spawn or milt. There are found also in Asia Minor, in rivers not communicating with the sea, little fishes like whitebait, differing from the small fry found near Cnidos but found under similar circumstances. Some writers actually aver that mullet all grow spontaneously. In this assertion they are mistaken, for the female of the fish is found provided with spawn, and the male with milt. However, there is a species of mullet that grows spontaneously out of mud and sand.

From the facts above enumerated it is quite proved that certain fish come spontaneously into existence, not being derived from eggs or from copulation. Such fish as are neither oviparous nor viviparous arise all from one of two sources, from mud, or from sand and from decayed matter that rises thence as a scum; for instance, the so-called froth of the small fry comes out of sandy ground. This fry is incapable of growth and of propagating its kind; after living for a while it dies away and another creature takes its place, and so, with short intervals excepted, it may be said to last the whole year through. At all events, it lasts from the autumn rising of Arcturus up to the spring-time. As a proof that these fish occasionally come out of the ground we have the fact that in cold weather they are not caught, and that they are caught in warm weather, obviously coming up out of the ground to catch the heat; also, when the fishermen use dredges and the ground is scraped up fairly often, the fishes appear in larger numbers and of superior quality. All other small fry are inferior in quality owing to rapidity of growth. The fry are found in sheltered and marshy districts, when after a spell of fine weather the ground is getting warmer, as, for instance, in the neighbourhood of Athens, at Salamis and near the tomb of Themistocles and at Marathon; for in these districts the froth is found. It appears, then, in such districts and during such weather, and occasionally appears after a heavy fall of rain in the froth that is thrown up by the falling rain, from which circumstance the substance derives its specific name. Foam is occasionally brought in on the surface of the sea in fair weather. (And in this, where it has formed on the surface, the so-called froth collects, as grubs swarm in manure; for which-reason this fry is often brought in from the open sea. The fish is at its best in quality and quantity in moist warm weather.)

The ordinary fry is the normal issue of parent fishes: the so-called gudgeon-fry of small insignificant gudgeon-like fish that burrow under the ground. From the Phaleric fry comes the membras, from the membras the trichis, from the trichis the trichias, and from one particular sort of fry, to wit from that found in the harbour of Athens, comes what is called the encrasicholus, or anchovy. There is another fry, derived from the maenis and the mullet.

The unfertile fry is watery and keeps only a short time, as has been stated, for at last only head and eyes are left. However, the fishermen of late have hit upon a method of transporting it to a distance, as when salted it keeps for a considerable time.

Part 16

Eels are not the issue of pairing, neither are they oviparous; nor was an eel ever found supplied with either milt or spawn, nor are they when cut open found to have within them passages for spawn or for eggs. In point of fact, this entire species of blooded animals proceeds neither from pair nor from the egg.

There can be no doubt that the case is so. For in some standing pools, after the water has been drained off and the mud has been dredged away, the eels appear again after a fall of rain. In time of drought they do not appear even in stagnant ponds, for the simple reason that their existence and sustenance is derived from rain-water.

There is no doubt, then, that they proceed neither from pairing nor from an egg. Some writers, however, are of opinion that they generate their kind, because in some eels little worms are found, from which they suppose that eels are derived. But this opinion is not founded on fact. Eels are derived from the so-called 'earth's guts' that grow spontaneously in mud and in humid ground; in fact, eels have at times been seen to emerge out of such earthworms, and on other occasions have been rendered visible when the earthworms were laid open by either scraping or cutting. Such earthworms are found both in the sea and in rivers, especially where there is decayed matter: in the sea in places where sea-weed abounds, and in rivers and marshes near to the edge; for it is near to the water's edge that sun-heat has its chief power and produces putrefaction. So much for the generation of the eel.

Part 17

Fish do not all bring forth their young at the same season nor all in like manner, neither is the period of gestation for all of the same duration.

Before pairing the males and females gather together in shoals; at the time for copulation and parturition they pair off. With some fish the time of gestation is not longer than thirty days, with others it is a lesser period; but with all it extends over a number of days divisible by seven. The longest period of gestation is that of the species which some call a marinus.

The sargue conceives during the month of Poseideon (or December), and carries its spawn for thirty days; and the species of mullet named by some the chelon, and the myxon, go with spawn at the same period and over the same length of time.

All fish suffer greatly during the period of gestation, and are in consequence very apt to be thrown up on shore at this time. In some cases they are driven frantic with pain and throw themselves on land. At all events they are throughout this time continually in motion until parturition is over (this being especially true of the mullet), and after parturition they are in repose. With many fish the time for parturition terminates on the appearance of grubs within the belly; for small living grubs get generated there and eat up the spawn.

With shoal fishes parturition takes place in the spring, and indeed, with most fishes, about the time of the spring equinox; with others it is at different times, in summer with some, and with others about the autumn equinox.

The first of shoal fishes to spawn is the atherine, and it spawns close to land; the last is the cephalus: and this is inferred from the fact that the brood of the atherine appears first of all and the brood of the cephalus last. The mullet also spawns early. The saupe spawns usually at the beginning of summer, but occasionally in the autumn. The aulopias, which some call the anthias, spawns in the summer. Next in order of spawning comes the chrysophrys or gilthead, the basse, the mormyrus, and in general such fish as are nicknamed 'runners'. Latest in order of the shoal fish come the red mullet and the coracine; these spawn in autumn. The red mullet spawns on mud, and consequently, as the mud continues cold for a long while, spawns late in the year. The coracine carries its spawn for a long time; but, as it lives usually on rocky ground, it goes to a distance and spawns in places abounding in seaweed, at a period later than the red mullet. The maenis spawns about the winter solstice. Of the others, such as are pelagic spawn for the most part in summer; which fact is proved by their not being caught by fishermen during this period.

Of ordinary fishes the most prolific is the sprat; of cartilaginous fishes, the fishing-frog. Specimens, however, of the fishing-frog are rare from the facility with which the young are destroyed, as the female lays her spawn all in a lump close in to shore. As a rule, cartilaginous fish are less prolific than other fish owing to their being viviparous; and their young by reason of their size have a better chance of escaping destruction.

The so-called needle-fish (or pipe-fish) is late in spawning, and the greater portion of them are burst asunder by the eggs before spawning; and the eggs are not so many in number as large in size. The young fish cluster round the parent like so many young spiders, for the fish spawns on to herself; and, if any one touch the young, they swim away. The atherine spawns by rubbing its belly against the sand.

Tunny fish also burst asunder by reason of their fat. They live for two years; and the fishermen infer this age from the circumstance that once when there was a failure of the young tunny fish for a year there was a failure of the full-grown tunny the next summer. They are of opinion that the tunny is a fish a year older than the pelamyd. The tunny and the mackerel pair about the close of the month of Elaphebolion, and spawn about the commencement of the month of Hecatombaeon; they deposit their spawn in a sort of bag. The growth of the young tunny is rapid. After the females have spawned in the Euxine, there comes from the egg what some call scordylae, but what the Byzantines nickname the 'auxids' or 'growers', from their growing to a considerable size in a few days; these fish go out of the Pontus in autumn along with the young tunnies, and enter Pontus in the spring as pelamyds. Fishes as a rule take on growth with rapidity, but this is peculiarly the case with all species of fish found in the Pontus; the growth, for instance, of the amia-tunny is quite visible from day to day.

To resume, we must bear in mind that the same fish in the same localities have not the same season for pairing, for conception, for parturition, or for favouring weather. The coracine, for instance, in some places spawns about wheat-harvest. The statements here given pretend only to give the results of general observation.

The conger also spawns, but the fact is not equally obvious in all localities, nor is the spawn plainly visible owing to the fat of the fish; for the spawn is lanky in shape as it is with serpents. However, if it be put on the fire it shows its nature; for the fat evaporates and melts, while the eggs dance about and explode with a crack. Further, if you touch the substances and rub them with your fingers, the fat feels smooth and the egg rough. Some congers are provided with fat but not with any spawn, others are unprovided with fat but have egg-spawn as here described.

Part 18

We have, then, treated pretty fully of the animals that fly in the air or swim in the water, and of such of those that walk on dry land as are oviparous, to wit of their pairing, conception, and the like phenomena; it now remains to treat of the same phenomena in connexion with viviparous land animals and with man.

The statements made in regard to the pairing of the sexes apply partly to the particular kinds of animal and partly to all in general. It is common to all animals to be most excited by the desire of one sex for the other and by the pleasure derived from copulation. The female is most cross-tempered just after parturition, the male during the time of pairing; for instance, stallions at this period bite one another, throw their riders, and chase them. Wild boars, though usually enfeebled at this time as the result of copulation, are now unusually fierce, and fight with one another in an extraordinary way, clothing themselves with defensive armour, or in other words deliberately thickening their hide by rubbing against trees or by coating themselves repeatedly all over with mud and then drying themselves in the sun. They drive one another away from the swine pastures, and fight with such fury that very often both combatants succumb. The case is similar with bulls, rams, and he-goats; for, though at ordinary times they herd together, at breeding time they hold aloof from and quarrel with one another. The male camel also is cross-tempered at pairing time if either a man or a camel comes near him; as for a horse, a camel is ready to fight him at any time. It is the same with wild animals. The bear, the wolf, and the lion are all at this time ferocious towards such as come in their way, but the males of these animals are less given to fight with one another from the fact that they are at no time gregarious. The she-bear is fierce after cubbing, and the bitch after pupping.

Male elephants get savage about pairing time, and for this reason it is stated that men who have charge of elephants in India never allow the males to have intercourse with the females; on the ground that the males go wild at this time and turn topsy-turvy the dwellings of their keepers, lightly constructed as they are, and commit all kinds of havoc. They also state that abundance of food has a tendency to tame the males. They further introduce other elephants amongst the wild ones, and punish and break them in by setting on the new-comers to chastise the others.

Animals that pair frequently and not at a single specific season, as for instance animals domesticated by man, such as swine and dogs, are found to indulge in such freaks to a lesser degree owing to the frequency of their sexual intercourse.

Of female animals the mare is the most sexually wanton, and next in order comes the cow. In fact, the mare is said to go a-horsing; and the term derived from the habits of this one animal serves as a term of abuse applicable to such females of the human species as are unbridled in the way of sexual appetite. This is the common phenomenon as observed in the sow when she is said to go a-boaring. The mare is said also about this time to get wind-impregnated if not impregnated by the stallion, and for this reason in Crete they never remove the stallion from the mares; for when the mare gets into this condition she runs away from all other horses. The mares under these circumstances fly invariably either northwards or southwards, and never towards either east or west. When this complaint is on them they allow no one to approach, until either they are exhausted with fatigue or have reached the sea. Under either of these circumstances they discharge a certain substance 'hippomanes', the title given to a growth on a new-born foal; this resembles the sow-virus, and is in great request amongst women who deal in drugs and potions. About horsing time the mares huddle closer together, are continually switching their tails, their neigh is abnormal in sound, and from the sexual organ there flows a liquid resembling genital sperm, but much thinner than the sperm of the male. It is this substance that some call hippomanes, instead of the growth found on the foal; they say it is extremely difficult to get as it oozes out only in small drops at a time. Mares also, when in heat, discharge urine frequently, and frisk with one another. Such are the phenomena connected with the horse.

Cows go a-bulling; and so completely are they under the influence of the sexual excitement that the herdsmen have no control over them and cannot catch hold of them in the fields. Mares and kine alike, when in heat, indicate the fact by the upraising of their genital organs, and by continually voiding urine. Further, kine mount the bulls, follow them about; and keep standing beside them. The younger females both with horses and oxen are the first to get in heat; and their sexual appetites are all the keener if the weather warm and their bodily condition be healthy. Mares, when clipped of their coat, have the sexual feeling checked, and assume a downcast drooping appearance. The stallion recognizes by the scent the mares that form his company, even though they have been together only a few days before breeding time: if they get mixed up with other mares, the stallion bites and drives away the interlopers. He feeds apart, accompanied by his own troop of mares. Each stallion has assigned to him about thirty mares or even somewhat more; when a strange stallion approaches, he huddles his mares into a close ring, runs round them, then advances to the encounter of the newcomer; if one of the mares make a movement, he bites her and drives her back. The bull in breeding time begins to graze with the cows, and fights with other bulls (having hitherto grazed with them), which is termed by graziers 'herd-spurning'. Often in Epirus a bull disappears for three months together. In a general way one may state that of male animals either none or few herd with their respective females before breeding time; but they keep separate after reaching maturity, and the two sexes feed apart. Sows, when they are moved by sexual desire, or are, as it is called, a-boaring, will attack even human beings.

With bitches the same sexual condition is termed 'getting into heat'. The sexual organ rises at this time, and there is a moisture about the parts. Mares drip with a white liquid at this season.

Female animals are subject to menstrual discharges, but never in such-abundance as is the female of the human species. With ewes and she-goats there are signs of menstruation in breeding time, just before the for submitting to the male; after copulation also the signs are manifest, and then cease for an interval until the period of parturition arrives; the process then supervenes, and it is by this supervention that the shepherd knows that such and such an ewe is about to bring forth. After parturition comes copious menstruation, not at first much tinged with blood, but deeply dyed with it by and by. With the cow, the she ass, and the mare, the discharge is more copious actually, owing to their greater bulk, but proportionally to the greater bulk it is far less copious. The cow, for instance, when in heat, exhibits a small discharge to the extent of a quarter of a pint of liquid or a little less; and the time when this discharge takes place is the best time for her to be covered by the bull. Of all quadrupeds the mare is the most easily delivered of its young, exhibits the least amount of discharge after parturition, and emits the least amount of blood; that is to say, of all animals in proportion to size. With kine and mares menstruation usually manifests itself at intervals of two, four, and six months; but, unless one be constantly attending to and thoroughly acquainted with such animals, it is difficult to verify the circumstance, and the result is that many people are under the belief that the process never takes place with these animals at all.

With mules menstruation never takes place, but the urine of the female is thicker than the urine of the male. As a general rule the discharge from the bladder in the case of quadrupeds is thicker than it is in the human species, and this discharge with ewes and she-goats is thicker than with rams and he-goats; but the urine of the jackass is thicker than the urine of the she-ass, and the urine of the bull is more pungent than the urine of the cow. After parturition the urine of all quadrupeds becomes thicker, especially with such animals as exhibit comparatively slight discharges. At breeding time the milk become purulent, but after parturition it becomes wholesome. During pregnancy ewes and she-goats get fatter and eat more; as is also the case with cows, and, indeed, with the females of all quadrupeds.

In general the sexual appetites of animals are keenest in spring-time; the time of pairing, however, is not the same for all, but is adapted so as to ensure the rearing of the young at a convenient season.

Domesticated swine carry their young for four months, and bring forth a litter of twenty at the utmost; and, by the way, if the litter be exceedingly numerous they cannot rear all the young. As the sow grows old she continues to bear, but grows indifferent to the boar; she conceives after a single copulation, but they have to put the boar to her repeatedly owing to her dropping after intercourse what is called the sow-virus. This incident befalls all sows, but some of them discharge the genital sperm as well. During conception any one of the litter that gets injured or dwarfed is called an afterpig or scut: such injury may occur at any part of the womb. After littering the mother offers the foremost teat to the first-born. When the sow is in heat, she must not at once be put to the boar, but only after she lets her lugs drop, for otherwise she is apt to get into heat again; if she be put to the boar when in full condition of heat, one copulation, as has been said, is sufficient. It is as well to supply the boar at the period of copulation with barley, and the sow at the time of parturition with boiled barley. Some swine give fine litters only at the beginning, with others the litters improve as the mothers grow in age and size. It is said that a sow, if she have one of her eyes knocked out, is almost sure to die soon afterwards. Swine for the most part live for fifteen years, but some fall little short of the twenty.

Part 19

Ewes conceive after three or four copulations with the ram. If rain falls after intercourse, the ram impregnates the ewe again; and it is the same with the she-goat. The ewe bears usually two lambs, sometimes three or four. Both ewe and she-goat carry their young for five months; consequently wherever a district is sunny and the animals are used to comfort and well fed, they bear twice in the year. The goat lives for eight years and the sheep for ten, but in most cases not so long; the bellwether, however, lives to fifteen years. In every flock they train one of the rams for bellwether. When he is called on by name by the shepherd, he takes the lead of the flock: and to this duty the creature is trained from its earliest years. Sheep in Ethiopia live for twelve or thirteen years, goats for ten or eleven. In the case of the sheep and the goat the two sexes have intercourse all their lives long.

Twins with sheep and goats may be due to richness of pasturage, or to the fact that either the ram or the he-goat is a twin-begetter or that the ewe or the she-goat is a twin-bearer. Of these animals some give birth to males and others to females; and the difference in this respect depends on the waters they drink and also on the sires. And if they submit to the male when north winds are blowing, they are apt to bear males; if when south winds are blowing, females. Such as bear females may get to bear males, due regard being paid to their looking northwards when put to the male. Ewes accustomed to be put to the ram early will refuse him if he attempt to mount them late. Lambs are born white and black according as white or black veins are under the ram's tongue; the lambs are white if the veins are white, and black if the veins are black, and white and black if the veins are white and black; and red if the veins are red. The females that drink salted waters are the first to take the male; the water should be salted before and after parturition, and again in the springtime. With goats the shepherds appoint no bellwether, as the animal is not capable of repose but frisky and apt to ramble. If at the appointed season the elders of the flock are eager for intercourse, the shepherds say that it bodes well for the flock; if the younger ones, that the flock is going to be bad.

Part 20

Of dogs there are several breeds. Of these the Laconian hound of either sex is fit for breeding purposes when eight months old: at about the same age some dogs lift the leg when voiding urine. The bitch conceives with one lining; this is clearly seen in the case where a dog contrives to line a bitch by stealth, as they impregnate after mounting only once. The Laconian bitch carries her young the sixth part of a year or sixty days: or more by one, two, or three, or less by one; the pups are blind for twelve days after birth. After pupping, the bitch gets in heat again in six months, but not before. Some bitches carry their young for the fifth part of the year or for seventy-two days; and their pups are blind for fourteen days. Other bitches carry their young for a quarter of a year or for three whole months; and the whelps of these are blind for seventeen days. The bitch appears go in heat for the same length of time. Menstruation continues for seven days, and a swelling of the genital organ occurs simultaneously; it is not during this period that the bitch is disposed to submit to the dog, but in the seven days that follow. The bitch as a rule goes in heat for fourteen days, but occasionally for sixteen. The birth-discharge occurs simultaneously with the delivery of the whelps, and the substance of it is thick and mucous. (The falling-off in bulk on the part of the mother is not so great as might have been inferred from the size of her frame.) The bitch is usually supplied with milk five days before parturition; some seven days previously, some four; and the milk is serviceable immediately after birth. The Laconian bitch is supplied with milk thirty days after lining. The milk at first is thickish, but gets thinner by degrees; with the bitch the milk is thicker than with the female of any other animal excepting the sow and the hare. When the bitch arrives at full growth an indication is given of her capacity for the male; that is to say, just as occurs in the female of the human species, a swelling takes place in the teats of the breasts, and the breasts take on gristle. This incident, however, it is difficult for any but an expert to detect, as the part that gives the indication is inconsiderable. The preceding statements relate to the female, and not one of them to the male. The male as a rule lifts his leg to void urine when six months old; some at a later period, when eight months old, some before they reach six months. In a general way one may put it that they do so when they are out of puppyhood. The bitch squats down when she voids urine; it is a rare exception that she lifts the leg to do so. The bitch bears twelve pups at the most, but usually five or six; occasionally a bitch will bear one only. The bitch of the Laconian breed generally bears eight. The two sexes have intercourse with each other at all periods of life. A very remarkable phenomenon is observed in the case of the Laconian hound: in other words, he is found to be more vigorous in commerce with the female after being hard-worked than when allowed to live idle.

The dog of the Laconian breed lives ten years, and the bitch twelve. The bitch of other breeds usually lives for fourteen or fifteen years, but some live to twenty; and for this reason certain critics consider that Homer did well in representing the dog of Ulysses as having died in his twentieth year. With the Laconian hound, owing to the hardships to which the male is put, he is less long-lived than the female; with other breeds the distinction as to longevity is not very apparent, though as a general rule the male is the longer-lived.

The dog sheds no teeth except the so-called 'canines'; these a dog of either sex sheds when four months old. As they shed these only, many people are in doubt as to the fact, and some people, owing to their shedding but two and its being hard to hit upon the time when they do so, fancy that the animal sheds no teeth at all; others, after observing the shedding of two, come to the conclusion that the creature sheds the rest in due turn. Men discern the age of a dog by inspection of its teeth; with young dogs the teeth are white and sharp pointed, with old dogs black and blunted.

Part 21

The bull impregnates the cow at a single mount, and mounts with such vigour as to weigh down the cow; if his effort be unsuccessful, the cow must be allowed an interval of twenty days before being again submitted. Bulls of mature age decline to mount the same cow several times on one day, except, by the way, at considerable intervals. Young bulls by reason of their vigour are enabled to mount the same cow several times in one day, and a good many cows besides. The bull is the least salacious of male animals.... The victor among the bulls is the one that mounts the females; when he gets exhausted by his amorous efforts, his beaten antagonist sets on him and very often gets the better of the conflict. The bull and the cow are about a year old when it is possible for them to have commerce with chance of offspring: as a rule, however, they are about twenty months old, but it is universally allowed that they are capable in this respect at the age of two years. The cow goes with calf for nine months, and she calves in the tenth month; some maintain that they go in calf for ten months, to the very day. A calf delivered before the times here specified is an abortion and never lives, however little premature its birth may have been, as its hooves are weak and imperfect. The cow as a rule bears but one calf, very seldom two; she submits to the bull and bears as long as she lives.

Cows live for about fifteen years, and the bulls too, if they have been castrated; but some live for twenty years or even more, if their bodily constitutions be sound. The herdsmen tame the castrated bulls, and give them an office in the herd analogous to the office of the bell-wether in a flock; and these bulls live to an exceptionally advanced age, owing to their exemption from hardship and to their browsing on pasture of good quality. The bull is in fullest vigour when five years old, which leads the critics to commend Homer for applying to the bull the epithets of 'five-year-old', or 'of nine seasons', which epithets are alike in meaning. The ox sheds his teeth at the age of two years, not all together but just as the horse sheds his. When the animal suffers from podagra it does not shed the hoof, but is subject to a painful swelling in the feet. The milk of the cow is serviceable after parturition, and before parturition there is no milk at all. The milk that first presents itself becomes as hard as stone when it clots; this result ensues unless it be previously diluted with water. Oxen younger than a year old do not copulate unless under circumstances of an unnatural and portentous kind: instances have been recorded of copulation in both sexes at the age of four months. Kine in general begin to submit to the male about the month of Thargelion or of Scirophorion; some, however, are capable of conception right on to the autumn. When kine in large numbers receive the bull and conceive, it is looked upon as prognostic of rain and stormy weather. Kine herd together like mares, but in lesser degree.

Part 22

In the case of horses, the stallion and the mare are first fitted for breeding purposes when two years old. Instances, however, of such early maturity are rare, and their young are exceptionally small and weak; the ordinary age for sexual maturity is three years, and from that age to twenty the two sexes go on improving in the quality of their offspring. The mare carries her foal for eleven months, and casts it in the twelfth. It is not a fixed number of days that the stallion takes to impregnate the mare; it may be one, two, three, or more. An ass in covering will impregnate more expeditiously than a stallion. The act of intercourse with horses is not laborious as it is with oxen. In both sexes the horse is the most salacious of animals next after the human species. The breeding faculties of the younger horses may be stimulated beyond their years if they be supplied with good feeding in abundance. The mare as a rule bears only one foal; occasionally she has two, but never more. A mare has been known to cast two mules; but such a circumstance was regarded as unnatural and portentous.

The horse then is first fitted for breeding purposes at the age of two and a half years, but achieves full sexual maturity when it has ceased to shed teeth, except it be naturally infertile; it must be added, however, that some horses have been known to impregnate the mare while the teeth were in process of shedding.

The horse has forty teeth. It sheds its first set of four, two from the upper jaw and two from the lower, when two and a half years old. After a year's interval, it sheds another set of four in like manner, and another set of four after yet another year's interval; after arriving at the age of four years and six months it sheds no more. An instance has occurred where a horse shed all his teeth at once, and another instance of a horse shedding all his teeth with his last set of four; but such instances are very rare. It consequently happens that a horse when four and a half years old is in excellent condition for breeding purposes.

The older horses, whether of the male or female, are the more generatively productive. Horses will cover mares from which they have been foaled and mares which they have begotten; and, indeed, a troop of horses is only considered perfect when such promiscuity of intercourse occurs. Scythians use pregnant mares for riding when the embryo has turned rather soon in the womb, and they assert that thereby the mothers have all the easier delivery. Quadrupeds as a rule lie down for parturition, and in consequence the young of them all come out of the womb sideways. The mare, however, when the time for parturition arrives, stands erect and in that posture casts its foal.

The horse in general lives for eighteen or twenty years; some horses live for twenty-five or even thirty, and if a horse be treated with extreme care, it may last on to the age of fifty years; a horse, however, when it reaches thirty years is regarded as exceptionally old. The mare lives usually for twenty-five years, though instances have occurred of their attaining the age of forty. The male is less long-lived than the female by reason of the sexual service he is called on to render; and horses that are reared in a private stable live longer than such as are reared in troops. The mare attains her full length and height at five years old, the stallion at six; in another six years the animal reaches its full bulk, and goes on improving until it is twenty years old. The female, then, reaches maturity more rapidly than the male, but in the womb the case is reversed, just as is observed in regard to the sexes of the human species; and the same phenomenon is observed in the case of all animals that bear several young.

The mare is said to suckle a mule-foal for six months, but not to allow its approach for any longer on account of the pain it is put to by the hard tugging of the young; an ordinary foal it allows to suck for a longer period.

Horse and mule are at their best after the shedding of the teeth. After they have shed them all, it is not easy to distinguish their age; hence they are said to carry their mark before the shedding, but not after. However, even after the shedding their age is pretty well recognized by the aid of the canines; for in the case of horses much ridden these teeth are worn away by attrition caused by the insertion of the bit; in the case of horses not ridden the teeth are large and detached, and in young horses they are sharp and small.

The male of the horse will breed at all seasons and during its whole life; the mare can take the horse all its life long, but is not thus ready to pair at all seasons unless it be held in check by a halter or some other compulsion be brought to bear. There is no fixed time at which intercourse of the two sexes cannot take place; and accordingly intercourse may chance to take place at a time that may render difficult the rearing of the future progeny. In a stable in Opus there was a stallion that used to serve mares when forty years old: his fore legs had to be lifted up for the operation.

Mares first take the horse in the spring-time. After a mare has foaled she does not get impregnated at once again, but only after a considerable interval; in fact, the foals will be all the better if the interval extend over four or five years. It is, at all events, absolutely necessary to allow an interval of one year, and for that period to let her lie fallow. A mare, then, breeds at intervals; a she-ass breeds on and on without intermission. Of mares some are absolutely sterile, others are capable of conception but incapable of bringing the foal to full term; it is said to be an indication of this condition in a mare, that her foal if dissected is found to have other kidney-shaped substances round about its kidneys, presenting the appearance of having four kidneys.

After parturition the mare at once swallows the after-birth, and bites off the growth, called the 'hippomanes', that is found on the forehead of the foal. This growth is somewhat smaller than a dried fig; and in shape is broad and round, and in colour black. If any bystander gets possession of it before the mare, and the mare gets a smell of it, she goes wild and frantic at the smell. And it is for this reason that venders of drugs and simples hold the substance in high request and include it among their stores.

If an ass cover a mare after the mare has been covered by a horse, the ass will destroy the previously formed embryo.

(Horse-trainers do not appoint a horse as leader to a troop, as herdsmen appoint a bull as leader to a herd, and for this reason that the horse is not steady but quick-tempered and skittish.)

Part 23

The ass of both sexes is capable of breeding, and sheds its first teeth at the age of two and a half years; it sheds its second teeth within six months, its third within another six months, and the fourth after the like interval. These fourth teeth are termed the gnomons or age-indicators.

A she-ass has been known to conceive when a year old, and the foal to be reared. After intercourse with the male it will discharge the genital sperm unless it be hindered, and for this reason it is usually beaten after such intercourse and chased about. It casts its young in the twelfth month. It usually bears but one foal, and that is its natural number, occasionally however it bears twins. The ass if it cover a mare destroys, as has been said, the embryo previously begotten by the horse; but, after the mare has been covered by the ass, the horse supervening will not spoil the embryo. The she-ass has milk in the tenth month of pregnancy. Seven days after casting a foal the she-ass submits to the male, and is almost sure to conceive if put to the male on this particular day; the same result, however, is quite possible later on. The she-ass will refuse to cast her foal with anyone looking on or in the daylight and just before foaling she has to be led away into a dark place. If the she-ass has had young before the shedding of the index-teeth, she will bear all her life through; but if not, then she will neither conceive nor bear for the rest of her days. The ass lives for more than thirty years, and the she-ass lives longer than the male.

When there is a cross between a horse and a she-ass or a jackass and a mare, there is much greater chance of a miscarriage than where the commerce is normal. The period for gestation in the case of a cross depends on the male, and is just what it would have been if the male had had commerce with a female of his own kind. In regard to size, looks, and vigour, the foal is more apt to resemble the mother than the sire. If such hybrid connexions be continued without intermittence, the female will soon go sterile; and for this reason trainers always allow of intervals between breeding times. A mare will not take the ass, nor a she-ass the horse, unless the ass or she-ass shall have been suckled by a mare; and for this reason trainers put foals of the she-ass under mares, which foals are technically spoken of as 'mare-suckled'. These asses, thus reared, mount the mares in the open pastures, mastering them by force as the stallions do.

Part 24

A mule is fitted for commerce with the female after the first shedding of its teeth, and at the age of seven will impregnate effectually; and where connexion has taken place with a mare, a 'hinny' has been known to be produced. After the seventh year it has no further intercourse with the female. A female mule has been known to be impregnated, but without the impregnation being followed up by parturition. In Syrophoenicia she-mules submit to the mule and bear young; but the breed, though it resembles the ordinary one, is different and specific. The hinny or stunted mule is foaled by a mare when she has gone sick during gestation, and corresponds to the dwarf in the human species and to the after-pig or scut in swine; and as is the case with dwarfs, the sexual organ of the hinny is abnormally large.

The mule lives for a number of years. There are on record cases of mules living to the age of eighty, as did one in Athens at the time of the building of the temple; this mule on account of its age was let go free, but continued to assist in dragging burdens, and would go side by side with the other draught-beasts and stimulate them to their work; and in consequence a public decree was passed forbidding any baker driving the creature away from his bread-tray. The she-mule grows old more slowly than the mule. Some assert that the she-mule menstruates by the act of voiding her urine, and that the mule owes the prematurity of his decay to his habit of smelling at the urine. So much for the modes of generation in connexion with these animals.

Part 25

Breeders and trainers can distinguish between young and old quadrupeds. If, when drawn back from the jaw, the skin at once goes back to its place, the animal is young; if it remains long wrinkled up, the animal is old.

Part 26

The camel carries its young for ten months, and bears but one at a time and never more; the young camel is removed from the mother when a year old. The animal lives for a long period, more than fifty years. It bears in spring-time, and gives milk until the time of the next conception. Its flesh and milk are exceptionally palatable. The milk is drunk mixed with water in the proportion of either two to one or three to one.

Part 27

The elephant of either sex is fitted for breeding before reaching the age of twenty. The female carries her young, according to some accounts, for two and a half years; according to others, for three years; and the discrepancy in the assigned periods is due to the fact that there are never human eyewitnesses to the commerce between the sexes. The female settles down on its rear to cast its young, and obviously suffers greatly during the process. The young one, immediately after birth, sucks the mother, not with its trunk but with the mouth; and can walk about and see distinctly the moment it is born.

Part 28

The wild sow submits to the boar at the beginning of winter, and in the spring-time retreats for parturition to a lair in some district inaccessible to intrusion, hemmed in with sheer cliffs and chasms and overshadowed by trees. The boar usually remains by the sow for thirty days. The number of the litter and the period gestation is the same as in the case of the domesticated congener. The sound of the grunt also is similar; only that the sow grunts continually, and the boar but seldom. Of the wild boars such as are castrated grow to the largest size and become fiercest: to which circumstance Homer alludes when he says:-

'He reared against him a wild castrated boar: it was not like a food-devouring brute, but like a forest-clad promontory.'

Wild boars become castrated owing to an itch befalling them in early life in the region of the testicles, and the castration is superinduced by their rubbing themselves against the trunks of trees.

Part 29

The hind, as has been stated, submits to the stag as a rule only under compulsion, as she is unable to endure the male often owing to the rigidity of the penis. However, they do occasionally submit to the stag as the ewe submits ram; and when they are in heat the hinds avoid one another. The stag is not constant to one particular hind, but after a while quits one and mates with others. The breeding time is after the rising of Arcturus, during the months of Boedromion and Maimacterion. The period of gestation lasts for eight months. Conception comes on a few days after intercourse; and a number of hinds can be impregnated by a single male. The hind, as a rule, bears but one fawn, although instances have been known of her casting two. Out of dread of wild beasts she casts her young by the side of the high-road. The young fawn grows with rapidity. Menstruation occurs at no other time with the hind; it takes place only after parturition, and the substance is phlegm-like.

The hind leads the fawn to her lair; this is her place of refuge, a cave with a single inlet, inside which she shelters herself against attack.

Fabulous stories are told concerning the longevity of the animal, but the stories have never been verified, and the brevity of the period of gestation and the rapidity of growth in the fawn would not lead one to attribute extreme longevity to this creature.

In the mountain called Elaphoeis or Deer Mountain, which is in Arginussa in Asia Minor-the place, by the way, where Alcibiades was assassinated-all the hinds have the ear split, so that, if they stray to a distance, they can be recognized by this mark; and the embryo actually has the mark while yet in the womb of the mother.

The hind has four teats like the cow. After the hinds have become pregnant, the males all segregate one by one, and in consequence of the violence of their sexual passions they keep each one to himself, dig a hole in the ground, and bellow from time to time; in all these particulars they resemble the goat, and their foreheads from getting wetted become black, as is also the case with the goat. In this way they pass the time until the rain falls, after which time they turn to pasture. The animal acts in this way owing to its sexual wantonness and also to its obesity; for in summer-time it becomes so exceptionally fat as to be unable to run: in fact at this period they can be overtaken by the hunters that pursue them on foot in the second or third run; and, by the way, in consequence of the heat of the weather and their getting out of breath they always make for water in their runs. In the rutting season, the flesh of the deer is unsavoury and rank, like the flesh of the he-goat. In winter-time the deer becomes thin and weak, but towards the approach of the spring he is at his best for running. When on the run the deer keeps pausing from time to time, and waits until his pursuer draws upon him, whereupon he starts off again. This habit appears due to some internal pain: at all events, the gut is so slender and weak that, if you strike the animal ever so softly, it is apt to break asunder, though the hide of the animal remains sound and uninjured.

Part 30

Bears, as has been previously stated, do not copulate with the male mounting the back of the female, but with the female lying down under the male. The she-bear goes with young for thirty days. She brings forth sometimes one cub, sometimes two cubs, and at most five. Of all animals the newly born cub of the she bear is the smallest in proportion to the size of the mother; that is to say, it is larger than a mouse but smaller than a weasel. It is also smooth and blind, and its legs and most of its organs are as yet inarticulate. Pairing takes Place in the month of Elaphebolion, and parturition about the time for retiring into winter quarters; about this time the bear and the she-bear are at the fattest. After the she-bear has reared her young, she comes out of her winter lair in the third month, when it is already spring. The female porcupine, by the way, hibernates and goes with young the same number of days as the she-bear, and in all respects as to parturition resembles this animal. When a she-bear is with young, it is a very hard task to catch her.

Part 31

It has already been stated that the lion and lioness copulate rearwards, and that these animals are opisthuretic. They do not copulate nor bring forth at all seasons indiscriminately, but once in the year only. The lioness brings forth in the spring, generally two cubs at a time, and six at the very most; but sometimes only one. The story about the lioness discharging her womb in the act of parturition is a pure fable, and was merely invented to account for the scarcity of the animal; for the animal is, as is well known, a rare animal, and is not found in many countries. In fact, in the whole of Europe it is only found in the strip between the rivers Achelous and Nessus. The cubs of the lioness when newly born are exceedingly small, and can scarcely walk when two months old. The Syrian lion bears cubs five times: five cubs at the first litter, then four, then three, then two, and lastly one; after this the lioness ceases to bear for the rest of her days. The lioness has no mane, but this appendage is peculiar to the lion. The lion sheds only the four so-called canines, two in the upper jaw and two in the lower; and it sheds them when it is six months old.

Part 32

The hyena in colour resembles the wolf, but is more shaggy, and is furnished with a mane running all along the spine. What is recounted concerning its genital organs, to the effect that every hyena is furnished with the organ both of the male and the female, is untrue. The fact is that the sexual organ of the male hyena resembles the same organ in the wolf and in the dog; the part resembling the female genital organ lies underneath the tail, and does to some extent resemble the female organ, but it is unprovided with duct or passage, and the passage for the residuum comes underneath it. The female hyena has the part that resembles the organ of the male, and, as in the case of the male, has it underneath her tail, unprovided with duct or passage; and after it the passage for the residuum, and underneath this the true female genital organ. The female hyena has a womb, like all other female animals of the same kind. It is an exceedingly rare circumstance to meet with a female hyena. At least a hunter said that out of eleven hyenas he had caught, only one was a female.

Part 33

Hares copulate in a rearward posture, as has been stated, for the animal is opisthuretic. They breed and bear at all seasons, superfetate during pregnancy, and bear young every month. They do not give birth to their young ones all together at one time, but bring them forth at intervals over as many days as the circumstances of each case may require. The female is supplied with milk before parturition; and after bearing submits immediately to the male, and is capable of conception while suckling her young. The milk in consistency resembles sow's milk. The young are born blind, as is the case with the greater part of the fissipeds or toed animals.

Part 34

The fox mounts the vixen in copulation, and the vixen bears young like the she-bear; in fact, her young ones are even more inarticulately formed. Before parturition she retires to sequestered places, so that it is a great rarity for a vixen to be caught while pregnant. After parturition she warms her young and gets them into shape by licking them. She bears four at most at a birth.

Part 35

The wolf resembles the dog in regard to the time of conception and parturition, the number of the litter, and the blindness of the newborn young. The sexes couple at one special period, and the female brings forth at the beginning of the summer. There is an account given of the parturition of the she-wolf that borders on the fabulous, to the effect that she confines her lying-in to within twelve particular days of the year. And they give the reason for this in the form of a myth, viz. that when they transported Leto in so many days from the land of the Hyperboreans to the island of Delos, she assumed the form of a she-wolf to escape the anger of Hera. Whether the account be correct or not has not yet been verified; I give it merely as it is currently told. There is no more of truth in the current statement that the she-wolf bears once and only once in her lifetime.

The cat and the ichneumon bear as many young as the dog, and live on the same food; they live about six years. The cubs of the panther are born blind like those of the wolf, and the female bears four at the most at one birth. The particulars of conception are the same for the thos, or civet, as for the dog; the cubs of the animal are born blind, and the female bears two, or three, or four at a birth. It is long in the body and low in stature; but not withstanding the shortness of its legs it is exceptionally fleet of foot, owing to the suppleness of its frame and its capacity for leaping.

Part 36

There is found in Syria a so-called mule. It is not the same as the cross between the horse and ass, but resembles it just as a wild ass resembles the domesticated congener, and derives its name from the resemblance. Like the wild ass, this wild mule is remarkable for its speed. The animals of this species interbreed with one another; and a proof of this statement may be gathered from the fact that a certain number of them were brought into Phrygia in the time of Pharnaces, the father of Pharnabazus, and the animal is there still. The number originally introduced was nine, and there are three there at the present day.

Part 37

The phenomena of generation in regard to the mouse are the most astonishing both for the number of the young and for the rapidity of recurrence in the births. On one occasion a she-mouse in a state of pregnancy was shut up by accident in a jar containing millet-seed, and after a little while the lid of the jar was removed and upwards of one hundred and twenty mice were found inside it.

The rate of propagation of field mice in country places, and the destruction that they cause, are beyond all telling. In many places their number is so incalculable that but very little of the corn-crop is left to the farmer; and so rapid is their mode of proceeding that sometimes a small farmer will one day observe that it is time for reaping, and on the following morning, when he takes his reapers afield, he finds his entire crop devoured. Their disappearance is unaccountable: in a few days not a mouse will there be to be seen. And yet in the time before these few days men fail to keep down their numbers by fumigating and unearthing them, or by regularly hunting them and turning in swine upon them; for pigs, by the way, turn up the mouse-holes by rooting with their snouts. Foxes also hunt them, and the wild ferrets in particular destroy them, but they make no way against the prolific qualities of the animal and the rapidity of its breeding. When they are super-abundant, nothing succeeds in thinning them down except the rain; but after heavy rains they disappear rapidly.

In a certain district of Persia when a female mouse is dissected the female embryos appear to be pregnant. Some people assert, and positively assert, that a female mouse by licking salt can become pregnant without the intervention of the male.

Mice in Egypt are covered with bristles like the hedgehog. There is also a different breed of mice that walk on their two hind-legs; their front legs are small and their hind-legs long; the breed is exceedingly numerous. There are many other breeds of mice than are here referred to.

Book VII

Part 1

AS to Man's growth, first within his mother's womb and afterward to old age, the course of nature, in so far as man is specially concerned, is after the following manner. And, by the way, the difference of male and female and of their respective organs has been dealt with heretofore. When twice seven years old, in the most of cases, the male begins to engender seed; and at the same time hair appears upon the pubes, in like manner, so Alcmaeon of Croton remarks, as plants first blossom and then seed. About the same time, the voice begins to alter, getting harsher and more uneven, neither shrill as formerly nor deep as afterward, nor yet of any even tone, but like an instrument whose strings are frayed and out of tune; and it is called, by way of by-word, the bleat of the billy-goat. Now this breaking of the voice is the more apparent in those who are making trial of their sexual powers; for in those who are prone to lustfulness the voice turns into the voice of a man, but not so in the continent. For if a lad strive diligently to hinder his voice from breaking, as some do of those who devote themselves to music, the voice lasts a long while unbroken and may even persist with little change. And the breasts swell and likewise the private parts, altering in size and shape. (And by the way, at this time of life those who try by friction to provoke emission of seed are apt to experience pain as well as voluptuous sensations.) At the same age in the female, the breasts swell and the so-called catamenia commence to flow; and this fluid resembles fresh blood. There is another discharge, a white one, by the way, which occurs in girls even at a very early age, more especially if their diet be largely of a fluid nature; and this malady causes arrest of growth and loss of flesh. In the majority of cases the catamenia are noticed by the time the breasts have grown to the height of two fingers' breadth. In girls, too, about this time the voice changes to a deeper note; for while in general the woman's voice is higher than the man's, so also the voices of girls are pitched in a higher key than the elder women's, just as the boy's are higher than the men's; and the girls' voices are shriller than the boys', and a maid's flute is tuned sharper than a lad's.

Girls of this age have much need of surveillance. For then in particular they feel a natural impulse to make usage of the sexual faculties that are developing in them; so that unless they guard against any further impulse beyond that inevitable one which their bodily development of itself supplies, even in the case of those who abstain altogether from passionate indulgence, they contract habits which are apt to continue into later life. For girls who give way to wantonness grow more and more wanton; and the same is true of boys, unless they be safeguarded from one temptation and another; for the passages become dilated and set up a local flux or running, and besides this the recollection of pleasure associated with former indulgence creates a longing for its repetition.

Some men are congenitally impotent owing to structural defect; and in like manner women also may suffer from congenital incapacity. Both men and women are liable to constitutional change, growing healthier or more sickly, or altering in the way of leanness, stoutness, and vigour; thus, after puberty some lads who were thin before grow stout and healthy, and the converse also happens; and the same is equally true of girls. For when in boy or girl the body is loaded with superfluous matter, then, when such superfluities are got rid of in the spermatic or catamenial discharge, their bodies improve in health and condition owing to the removal of what had acted as an impediment to health and proper nutrition; but in such as are of opposite habit their bodies become emaciated and out of health, for then the spermatic discharge in the one case and the catamenial flow in the other take place at the cost of natural healthy conditions.

Furthermore, in the case of maidens the condition of the breasts is diverse in different individuals, for they are sometimes quite big and sometimes little; and as a general rule their size depends on whether or not the body was burthened in childhood with superfluous material. For when the signs of womanhood are nigh but not come, the more there be of moisture the more will it cause the breasts to swell, even to the bursting point; and the result is that the breasts remain during after-life of the bulk that they then acquired. And among men, the breasts grow more conspicuous and more like to those of women, both in young men and old, when the individual temperament is moist and sleek and the reverse of sinewy, and all the more among the dark-complexioned than the fair.

At the outset and till the age of one and twenty the spermatic discharge is devoid of fecundity; afterwards it becomes fertile, but young men and women produce undersized and imperfect progeny, as is the case also with the common run of animals. Young women conceive readily, but, having conceived, their labour in childbed is apt to be difficult.

The frame fails of reaching its full development and ages quickly in men of intemperate lusts and in women who become mothers of many children; for it appears to be the case that growth ceases when the woman has given birth to three children. Women of a lascivious disposition grow more sedate and virtuous after they have borne several children.

After the age of twenty-one women are fully ripe for child-bearing, but men go on increasing in vigour. When the spermatic fluid is of a thin consistency it is infertile; when granular it is fertile and likely to produce male children, but when thin and unclotted it is apt to produce female offspring. And it is about this time of life that in men the beard makes its appearance.

Part 2

The onset of the catamenia in women takes place towards the end of the month; and on this account the wiseacres assert that the moon is feminine, because the discharge in women and the waning of the moon happen at one and the same time, and after the wane and the discharge both one and the other grow whole again. (In some women the catamenia occur regularly but sparsely every month, and more abundantly every third month.) With those in whom the ailment lasts but a little while, two days or three, recovery is easy; but where the duration is longer, the ailment is more troublesome. For women are ailing during these days; and sometimes the discharge is sudden and sometimes gradual, but in all cases alike there is bodily distress until the attack be over. In many cases at the commencement of the attack, when the discharge is about to appear, there occur spasms and rumbling noises within the womb until such time as the discharge manifests itself.

Under natural conditions it is after recovery from these symptoms that conception takes place in women, and women in whom the signs do not manifest themselves for the most part remain childless. But the rule is not without exception, for some conceive in spite of the absence of these symptoms; and these are cases in which a secretion accumulates, not in such a way as actually to issue forth, but in amount equal to the residuum left in the case of child-bearing women after the normal discharge has taken place. And some conceive while the signs are on but not afterwards, those namely in whom the womb closes up immediately after the discharge. In some cases the menses persist during pregnancy up to the very last; but the result in these cases is that the offspring are poor, and either fail to survive or grow up weakly.

In many cases, owing to excessive desire, arising either from youthful impetuosity or from lengthened abstinence, prolapsion of the womb takes place and the catamenia appear repeatedly, thrice in the month, until conception occurs; and then the womb withdraws upwards again to its proper place...

As we have remarked above, the discharge is wont to be more abundant in women than in the females of any other animals. In creatures that do not bring forth their young alive nothing of the sort manifests itself, this particular superfluity being converted into bodily substance; and by the way, in such animals the females are sometimes larger than the males; and moreover, the material is used up sometimes for scutes and sometimes for scales, and sometimes for the abundant covering of feathers, whereas in the vivipara possessed of limbs it is turned into hair and into bodily substance (for man alone among them is smooth-skinned), and into urine, for this excretion is in the majority of such animals thick and copious. Only in the case of women is the superfluity turned into a discharge instead of being utilized in these other ways.

There is something similar to be remarked of men: for in proportion to his size man emits more seminal fluid than any other animal (for which reason man is the smoothest of animals), especially such men as are of a moist habit and not over corpulent, and fair men in greater degree than dark. It is likewise with women; for in the stout, great part of the excretion goes to nourish the body. In the act of intercourse, women of a fair complexion discharge a more plentiful secretion than the dark; and furthermore, a watery and pungent diet conduces to this phenomenon.

Part 3

It is a sign of conception in women when the place is dry immediately after intercourse. If the lips of the orifice be smooth conception is difficult, for the matter slips off; and if they be thick it is also difficult. But if on digital examination the lips feel somewhat rough and adherent, and if they be likewise thin, then the chances are in favour of conception. Accordingly, if conception be desired, we must bring the parts into such a condition as we have just described; but if on the contrary we want to avoid conception then we must bring about a contrary disposition. Wherefore, since if the parts be smooth conception is prevented, some anoint that part of the womb on which the seed falls with oil of cedar, or with ointment of lead or with frankincense, commingled with olive oil. If the seed remain within for seven days then it is certain that conception has taken place; for it is during that period that what is known as effluxion takes place.

In most cases the menstrual discharge recurs for some time after conception has taken place, its duration being mostly thirty days in the case of a female and about forty days in the case of a male child. After parturition also it is common for the discharge to be withheld for an equal number of days, but not in all cases with equal exactitude. After conception, and when the above-mentioned days are past, the discharge no longer takes its natural course but finds its way to the breasts and turns to milk. The first appearance of milk in the breasts is scant in quantity and so to speak cobwebby or interspersed with little threads. And when conception has taken place, there is apt to be a sort of feeling in the region of the flanks, which in some cases quickly swell up a little, especially in thin persons, and also in the groin.

In the case of male children the first movement usually occurs on the right-hand side of the womb and about the fortieth day, but if the child be a female then on the left-hand side and about the ninetieth day. However, we must by no means assume this to be an accurate statement of fact, for there are many exceptions, in which the movement is manifested on the right-hand side though a female child be coming, and on the left-hand side though the infant be a male. And in short, these and all suchlike phenomena are usually subject to differences that may be summed up as differences of degree.

About this period the embryo begins to resolve into distinct parts, it having hitherto consisted of a fleshlike substance without distinction of parts.

What is called effluxion is a destruction of the embryo within the first week, while abortion occurs up to the fortieth day; and the greater number of such embryos as perish do so within the space of these forty days.

In the case of a male embryo aborted at the fortieth day, if it be placed in cold water it holds together in a sort of membrane, but if it be placed in any other fluid it dissolves and disappears. If the membrane be pulled to bits the embryo is revealed, as big as one of the large kind of ants; and all the limbs are plain to see, including the penis, and the eyes also, which as in other animals are of great size. But the female embryo, if it suffer abortion during the first three months, is as a rule found to be undifferentiated; if however it reach the fourth month it comes to be subdivided and quickly attains further differentiation. In short, while within the womb, the female infant accomplishes the whole development of its parts more slowly than the male, and more frequently than the man-child takes ten months to come to perfection. But after birth, the females pass more quickly than the males through youth and maturity and age; and this is especially true of those that bear many children, as indeed I have already said.

Part 4

When the womb has conceived the seed, straightway in the majority of cases it closes up until seven months are fulfilled; but in the eighth month it opens, and the embryo, if it be fertile, descends in the eighth month. But such embryos as are not fertile but are devoid of breath at eight months old, their mothers do not bring into the world by parturition at eight months, neither does the embryo descend within the womb at that period nor does the womb open. And it is a sign that the embryo is not capable of life if it be formed without the above-named circumstances taking place.

After conception women are prone to a feeling of heaviness in all parts of their bodies, and for instance they experience a sensation of darkness in front of the eyes and suffer also from headache. These symptoms appear sooner or later, sometimes as early as the tenth day, according as the patient be more or less burthened with superfluous humours. Nausea also and sickness affect the most of women, and especially such as those that we have just now mentioned, after the menstrual discharge has ceased and before it is yet turned in the direction of the breasts.

Moreover, some women suffer most at the beginning of their pregnancy and some at a later period when the embryo has had time to grow; and in some women it is a common occurrence to suffer from strangury towards the end of their time. As a general rule women who are pregnant of a male child escape comparatively easily and retain a comparatively healthy look, but it is otherwise with those whose infant is a female; for these latter look as a rule paler and suffer more pain, and in many cases they are subject to swellings of the legs and eruptions on the body. Nevertheless the rule is subject to exceptions.

Women in pregnancy are a prey to all sorts of longings and to rapid changes of mood, and some folks call this the 'ivy-sickness'; and with the mothers of female infants the longings are more acute, and they are less contented when they have got what they desired.

In a certain few cases the patient feels unusually well during pregnancy. The worst time of all is just when the child's hair is beginning to grow.

In pregnant women their own natural hair is inclined to grow thin and fall out, but on the other hand hair tends to grow on parts of the body where it was not wont to be. As a general rule, a man-child is more prone to movement within its mother's womb than a female child, and it is usually born sooner. And labour in the case of female children is apt to be protracted and sluggish, while in the case of male children it is acute and by a long way more difficult. Women who have connexion with their husbands shortly before childbirth are delivered all the more quickly. Occasionally women seem to be in the pains of labour though labour has not in fact commenced, what seemed like the commencement of labour being really the result of the fetus turning its head.

Now all other animals bring the time of pregnancy to an end in a uniform way; in other words, one single term of pregnancy is defined for each of them. But in the case of mankind alone of all animals the times are diverse; for pregnancy may be of seven months' duration, or of eight months or of nine, and still more commonly of ten months, while some few women go even into the eleventh month.

Children that come into the world before seven months can under no circumstances survive. The seven-months' children are the earliest that are capable of life, and most of them are weakly-for which reason, by the way, it is customary to swaddle them in wool,-and many of them are born with some of the orifices of the body imperforate, for instance the ears or the nostrils. But as they get bigger they become more perfectly developed, and many of them grow up.

In Egypt, and in some other places where the women are fruitful and are wont to bear and bring forth many children without difficulty, and where the children when born are capable of living even if they be born subject to deformity, in these places the eight-months' children live and are brought up, but in Greece it is only a few of them that survive while most perish. And this being the general experience, when such a child does happen to survive the mother is apt to think that it was not an eight months' child after all, but that she had conceived at an earlier period without being aware of it.

Women suffer most pain about the fourth and the eighth months, and if the fetus perishes in the fourth or in the eighth month the mother also succumbs as a general rule; so that not only do the eight-months' children not live, but when they die their mothers are in great danger of their own lives. In like manner children that are apparently born at a later term than eleven months are held to be in doubtful case; inasmuch as with them also the beginning of conception may have escaped the notice of the mother. What I mean to say is that often the womb gets filled with wind, and then when at a later period connexion and conception take place, they think that the former circumstance was the beginning of conception from the similarity of the symptoms that they experienced.

Such then are the differences between mankind and other animals in regard to the many various modes of completion of the term of pregnancy. Furthermore, some animals produce one and some produce many at a birth, but the human species does sometimes the one and sometimes the other. As a general rule and among most nations the women bear one child a birth; but frequently and in many lands they bear twins, as for instance in Egypt especially. Sometimes women bring forth three and even four children, and especially in certain parts of the world, as has already been stated. The largest number ever brought forth is five, and such an occurrence has been witnessed on several occasions. There was once upon a time a certain women who had twenty children at four births; each time she had five, and most of them grew up.

Now among other animals, if a pair of twins happen to be male and female they have as good a chance of surviving as though both had been males or both females; but among mankind very few twins survive if one happen to be a boy and the other a girl.

Of all animals the woman and the mare are most inclined to receive the commerce of the male during pregnancy; while all other animals when they are pregnant avoid the male, save those in which the phenomenon of superfetation occurs, such as the hare. Unlike that animal, the mare after once conceiving cannot be rendered pregnant again, but brings forth one foal only, at least as a general rule; in the human species cases of superfetation are rare, but they do happen now and then.

An embryo conceived some considerable time after a previous conception does not come to perfection, but gives rise to pain and causes the destruction of the earlier embryo; and, by the way, a case has been known to occur where owing to this destructive influence no less than twelve embryos conceived by superfetation have been discharged. But if the second conception take place at a short interval, then the mother bears that which was later conceived, and brings forth the two children like actual twins, as happened, according to the legend, in the case of Iphicles and Hercules. The following also is a striking example: a certain woman, having committed adultery, brought forth the one child resembling her husband and the other resembling the adulterous lover.

The case has also occurred where a woman, being pregnant of twins, has subsequently conceived a third child; and in course of time she brought forth the twins perfect and at full term, but the third a five-months' child; and this last died there and then. And in another case it happened that the woman was first delivered of a seven-months' child, and then of two which were of full term; and of these the first died and the other two survived.

Some also have been known to conceive while about to miscarry, and they have lost the one child and been delivered of the other.

If women while going with child cohabit after the eighth month the child is in most cases born covered over with a slimy fluid. Often also the child is found to be replete with food of which the mother had partaken.

Part 5

When women have partaken of salt in overabundance their children are apt to be born destitute of nails.

Milk that is produced earlier than the seventh month is unfit for use; but as soon as the child is fit to live the milk is fit to use. The first of the milk is saltish, as it is likewise with sheep. Most women are sensibly affected by wine during pregnancy, for if they partake of it they grow relaxed and debilitated.

The beginning of child-bearing in women and of the capacity to procreate in men, and the cessation of these functions in both cases, coincide in the one case with the emission of seed and in the other with the discharge of the catamenia: with this qualification that there is a lack of fertility at the commencement of these symptoms, and again towards their close when the emissions become scanty and weak. The age at which the sexual powers begin has been related already. As for their end, the menstrual discharges ceases in most women about their fortieth year; but with those in whom it goes on longer it lasts even to the fiftieth year, and women of that age have been known to bear children. But beyond that age there is no case on record.

Part 6

Men in most cases continue to be sexually competent until they are sixty years old, and if that limit be overpassed then until seventy years; and men have been actually known to procreate children at seventy years of age. With many men and many women it so happens that they are unable to produce children to one another, while they are able to do so in union with other individuals. The same thing happens with regard to the production of male and female offspring; for sometimes men and women in union with one another produce male children or female, as the case may be, but children of the opposite sex when otherwise mated. And they are apt to change in this respect with advancing age: for sometimes a husband and wife while they are young produce female children and in later life male children; and in other cases the very contrary occurs. And just the same thing is true in regard to the generative faculty: for some while young are childless, but have children when they grow older; and some have children to begin with, and later on no more.

There are certain women who conceive with difficulty, but if they do conceive, bring the child to maturity; while others again conceive readily, but are unable to bring the child to birth. Furthermore, some men and some women produce female offspring and some male, as for instance in the story of Hercules, who among all his two and seventy children is said to have begotten but one girl. Those women who are unable to conceive, save with the help of medical treatment or some other adventitious circumstance, are as a general rule apt to bear female children rather than male.

It is a common thing with men to be at first sexually competent and afterwards impotent, and then again to revert to their former powers.

From deformed parents come deformed children, lame from lame and blind from blind, and, speaking generally, children often inherit anything that is peculiar in their parents and are born with similar marks, such as pimples or scars. Such things have been known to be handed down through three generations; for instance, a certain man had a mark on his arm which his son did not possess, but his grandson had it in the same spot though not very distinct.

Such cases, however, are few; for the children of cripples are mostly sound, and there is no hard and fast rule regarding them. While children mostly resemble their parents or their ancestors, it sometimes happens that no such resemblance is to be traced. But parents may pass on resemblance after several generations, as in the case of the woman in Elis, who committed adultery with a negro; in this case it was not the woman's own daughter but the daughter's child that was a blackamoor.

As a rule the daughters have a tendency to take after the mother, and the boys after the father; but sometimes it is the other way, the boys taking after the mother and the girls after the father. And they may resemble both parents in particular features.

There have been known cases of twins that had no resemblance to one another, but they are alike as a general rule. There was once upon a time a woman who had intercourse with her husband a week after giving birth to a child and she conceived and bore a second child as like the first as any twin. Some women have a tendency to produce children that take after themselves, and others children that take after the husband; and this latter case is like that of the celebrated mare in Pharsalus, that got the name of the Honest Wife.

Part 7

In the emission of sperm there is a preliminary discharge of air, and the outflow is manifestly caused by a blast of air; for nothing is cast to a distance save by pneumatic pressure. After the seed reaches the womb and remains there for a while, a membrane forms around it; for when it happens to escape before it is distinctly formed, it looks like an egg enveloped in its membrane after removal of the eggshell; and the membrane is full of veins.

All animals whatsoever, whether they fly or swim or walk upon dry land, whether they bring forth their young alive or in the egg, develop in the same way: save only that some have the navel attached to the womb, namely the viviparous animals, and some have it attached to the egg, and some to both parts alike, as in a certain sort of fishes. And in some cases membranous envelopes surround the egg, and in other cases the chorion surrounds it. And first of all the animal develops within the innermost envelope, and then another membrane appears around the former one, which latter is for the most part attached to the womb, but is in part separated from it and contains fluid. In between is a watery or sanguineous fluid, which the women folk call the forewaters.

Part 8

All animals, or all such as have a navel, grow by the navel. And the navel is attached to the cotyledon in all such as possess cotyledons, and to the womb itself by a vein in all such as have the womb smooth. And as regards their shape within the womb, the four-footed animals all lie stretched out, and the footless animals lie on their sides, as for instance fishes; but two-legged animals lie in a bent position, as for instance birds; and human embryos lie bent, with nose between the knees and eyes upon the knees, and the ears free at the sides.

All animals alike have the head upwards to begin with; but as they grow and approach the term of egress from the womb they turn downwards, and birth in the natural course of things takes place in all animals head foremost; but in abnormal cases it may take place in a bent position, or feet foremost.

The young of quadrupeds when they are near their full time contain excrements, both liquid and in the form of solid lumps, the latter in the lower part of the bowel and the urine in the bladder.

In those animals that have cotyledons in the womb the cotyledons grow less as the embryo grows bigger, and at length they disappear altogether. The navel-string is a sheath wrapped about blood-vessels which have their origin in the womb, from the cotyledons in those animals which possess them and from a blood-vessel in those which do not. In the larger animals, such as the embryos of oxen, the vessels are four in number, and in smaller animals two; in the very little ones, such as fowls, one vessel only.

Of the four vessels that run into the embryo, two pass through the liver where the so-called gates or 'portae' are, running in the direction of the great vein, and the other two run in the direction of the aorta towards the point where it divides and becomes two vessels instead of one. Around each pair of blood-vessels are membranes, and surrounding these membranes is the navel-string itself, after the manner of a sheath. And as the embryo grows, the veins themselves tend more and more to dwindle in size. And also as the embryo matures it comes down into the hollow of the womb and is observed to move here, and sometimes rolls over in the vicinity of the groin.

Part 9

When women are in labour, their pains determine towards many divers parts of the body, and in most cases to one or other of the thighs. Those are the quickest to be delivered who experience severe pains in the region of the belly; and parturition is difficult in those who begin by suffering pain in the loins, and speedy when the pain is abdominal. If the child about to be born be a male, the preliminary flood is watery and pale in colour, but if a girl it is tinged with blood, though still watery. In some cases of labour these latter phenomena do not occur, either one way or the other.

In other animals parturition is unaccompanied by pain, and the dam is plainly seen to suffer but moderate inconvenience. In women, however, the pains are more severe, and this is especially the case in persons of sedentary habits, and in those who are weak-chested and short of breath. Labour is apt to be especially difficult if during the process the woman while exerting force with her breath fails to hold it in.

First of all, when the embryo starts to move and the membranes burst, there issues forth the watery flood; then afterwards comes the embryo, while the womb everts and the afterbirth comes out from within.

Part 10

The cutting of the navel-string, which is the nurse's duty, is a matter calling for no little care and skill. For not only in cases of difficult labour must she be able to render assistance with skillful hand, but she must also have her wits about her in all contingencies, and especially in the operation of tying the cord. For if the afterbirth have come away, the navel is ligatured off from the afterbirth with a woolen thread and is then cut above the ligature; and at the place where it has been tied it heals up, and the remaining portion drops off. (If the ligature come loose the child dies from loss of blood.) But if the afterbirth has not yet come away, but remains after the child itself is extruded, it is cut away within after the ligaturing of the cord.

It often happens that the child appears to have been born dead when it is merely weak, and when before the umbilical cord has been ligatured, the blood has run out into the cord and its surroundings. But experienced midwives have been known to squeeze back the blood into the child's body from the cord, and immediately the child that a moment before was bloodless came back to life again.

It is the natural rule, as we have mentioned above, for all animals to come into the world head foremost, and children, moreover, have their hands stretched out by their sides. And the child gives a cry and puts its hands up to its mouth as soon as it issues forth.

Moreover the child voids excrement sometimes at once, sometimes a little later, but in all cases during the first day; and this excrement is unduly copious in comparison with the size of the child; it is what the midwives call the meconium or 'poppy-juice'. In colour it resembles blood, extremely dark and pitch-like, but later on it becomes milky, for the child takes at once to the breast. Before birth the child makes no sound, even though in difficult labour it put forth its head while the rest of the body remains within.

In cases where flooding takes place rather before its time, it is apt to be followed by difficult parturition. But if discharge take place after birth in small quantity, and in cases where it only takes place at the beginning and does not continue till the fortieth day, then in such cases women make a better recovery and are the sooner ready to conceive again.

Until the child is forty days old it neither laughs nor weeps during waking hours, but of nights it sometimes does both; and for the most part it does not even notice being tickled, but passes most of its time in sleep. As it keeps on growing, it gets more and more wakeful; and moreover it shows signs of dreaming, though it is long afterwards before it remembers what it dreams.

In other animals there is no contrasting difference between one bone and another, but all are properly formed; but in children the front part of the head is soft and late of ossifying. And by the way, some animals are born with teeth, but children begin to cut their teeth in the seventh month; and the front teeth are the first to come through, sometimes the upper and sometimes the lower ones. And the warmer the nurses' milk so much the quicker are the children's teeth to come.

Part 11

After parturition and the cleansing flood the milk comes in plenty, and in some women it flows not only from the nipples but at divers parts of the breasts, and in some cases even from the armpits. And for some time afterwards there continue to be certain indurated parts of the breast called strangalides, or 'knots', which occur when it so happens that the moisture is not concocted, or when it finds no outlet but accumulates within. For the whole breast is so spongy that if a woman in drinking happen to swallow a hair, she gets a pain in her breast, which ailment is called 'trichia'; and the pain lasts till the hair either find its own way out or be sucked out with the milk. Women continue to have milk until their next conception; and then the milk stops coming and goes dry, alike in the human species and in the quadrupedal vivipara. So long as there is a flow of milk the menstrual purgations do not take place, at least as a general rule, though the discharge has been known to occur during the period of suckling. For, speaking generally, a determination of moisture does not take place at one and the same time in several directions; as for instance the menstrual purgations tend to be scanty in persons suffering from hemorrhoids. And in some women the like happens owing to their suffering from varices, when the fluids issue from the pelvic region before entering into the womb. And patients who during suppression of the menses happen to vomit blood are no whit the worse.

Part 12

Children are very commonly subject to convulsions, more especially such of them as are more than ordinarily well-nourished on rich or unusually plentiful milk from a stout nurse. Wine is bad for infants, in that it tends to excite this malady, and red wine is worse than white, especially when taken undiluted; and most things that tend to induce flatulency are also bad, and constipation too is prejudicial. The majority of deaths in infancy occur before the child is a week old, hence it is customary to name the child at that age, from a belief that it has now a better chance of survival. This malady is worst at the full of the moon; and by the way, it is a dangerous symptom when the spasms begin in the child's back.

Book VIII

Part 1

WE have now discussed the physical characteristics of animals and their methods of generation. Their habits and their modes of living vary according to their character and their food.

In the great majority of animals there are traces of psychical qualities or attitudes, which qualities are more markedly differentiated in the case of human beings. For just as we pointed out resemblances in the physical organs, so in a number of animals we observe gentleness or fierceness, mildness or cross temper, courage, or timidity, fear or confidence, high spirit or low cunning, and, with regard to intelligence, something equivalent to sagacity. Some of these qualities in man, as compared with the corresponding qualities in animals, differ only quantitatively: that is to say, a man has more or less of this quality, and an animal has more or less of some other; other qualities in man are represented by analogous and not identical qualities: for instance, just as in man we find knowledge, wisdom, and sagacity, so in certain animals there exists some other natural potentiality akin to these. The truth of this statement will be the more clearly apprehended if we have regard to the phenomena of childhood: for in children may be observed the traces and seeds of what will one day be settled psychological habits, though psychologically a child hardly differs for the time being from an animal; so that one is quite justified in saying that, as regards man and animals, certain psychical qualities are identical with one another, whilst others resemble, and others are analogous to, each other.

Nature proceeds little by little from things lifeless to animal life in such a way that it is impossible to determine the exact line of demarcation, nor on which side thereof an intermediate form should lie. Thus, next after lifeless things in the upward scale comes the plant, and of plants one will differ from another as to its amount of apparent vitality; and, in a word, the whole genus of plants, whilst it is devoid of life as compared with an animal, is endowed with life as compared with other corporeal entities. Indeed, as we just remarked, there is observed in plants a continuous scale of ascent towards the animal. So, in the sea, there are certain objects concerning which one would be at a loss to determine whether they be animal or vegetable. For instance, certain of these objects are fairly rooted, and in several cases perish if detached; thus the pinna is rooted to a particular spot, and the solen (or razor-shell) cannot survive withdrawal from its burrow. Indeed, broadly speaking, the entire genus of testaceans have a resemblance to vegetables, if they be contrasted with such animals as are capable of progression.

In regard to sensibility, some animals give no indication whatsoever of it, whilst others indicate it but indistinctly. Further, the substance of some of these intermediate creatures is fleshlike, as is the case with the so-called tethya (or ascidians) and the acalephae (or sea-anemones); but the sponge is in every respect like a vegetable. And so throughout the entire animal scale there is a graduated differentiation in amount of vitality and in capacity for motion.

A similar statement holds good with regard to habits of life. Thus of plants that spring from seed the one function seems to be the reproduction of their own particular species, and the sphere of action with certain animals is similarly limited. The faculty of reproduction, then, is common to all alike. If sensibility be superadded, then their lives will differ from one another in respect to sexual intercourse through the varying amount of pleasure derived therefrom, and also in regard to modes of parturition and ways of rearing their young. Some animals, like plants, simply procreate their own species at definite seasons; other animals busy themselves also in procuring food for their young, and after they are reared quit them and have no further dealings with them; other animals are more intelligent and endowed with memory, and they live with their offspring for a longer period and on a more social footing.

The life of animals, then, may be divided into two acts-procreation and feeding; for on these two acts all their interests and life concentrate. Their food depends chiefly on the substance of which they are severally constituted; for the source of their growth in all cases will be this substance. And whatsoever is in conformity with nature is pleasant, and all animals pursue pleasure in keeping with their nature.

Part 2

Animals are also differentiated locally: that is to say, some live upon dry land, while others live in the water. And this differentiation may be interpreted in two different ways. Thus, some animals are termed terrestrial as inhaling air, and others aquatic as taking in water; and there are others which do not actually take in these elements, but nevertheless are constitutionally adapted to the cooling influence, so far as is needful to them, of one element or the other, and hence are called terrestrial or aquatic though they neither breathe air nor take in water. Again, other animals are so called from their finding their food and fixing their habitat on land or in water: for many animals, although they inhale air and breed on land, yet derive their food from the water, and live in water for the greater part of their lives; and these are the only animals to which as living in and on two elements the term 'amphibious' is applicable. There is no animal taking in water that is terrestrial or aerial or that derives its food from the land, whereas of the great number of land animals inhaling air many get their food from the water; moreover some are so peculiarly organized that if they be shut off altogether from the water they cannot possibly live, as for instance, the so-called sea-turtle, the crocodile, the hippopotamus, the seal, and some of the smaller creatures, such as the fresh-water tortoise and the frog: now all these animals choke or drown if they do not from time to time breathe atmospheric air: they breed and rear their young on dry land, or near the land, but they pass their lives in water.

But the dolphin is equipped in the most remarkable way of all animals: the dolphin and other similar aquatic animals, including the other cetaceans which resemble it; that is to say, the whale, and all the other creatures that are furnished with a blow-hole. One can hardly allow that such an animal is terrestrial and terrestrial only, or aquatic and aquatic only, if by terrestrial we mean an animal that inhales air, and if by aquatic we mean an animal that takes in water. For the fact is the dolphin performs both these processes: he takes in water and discharges it by his blow-hole, and he also inhales air into his lungs; for, by the way, the creature is furnished with this organ and respires thereby, and accordingly, when caught in the nets, he is quickly suffocated for lack of air. He can also live for a considerable while out of the water, but all this while he keeps up a dull moaning sound corresponding to the noise made by air-breathing animals in general; furthermore, when sleeping, the animal keeps his nose above water, and he does so that he may breathe the air. Now it would be unreasonable to assign one and the same class of animals to both categories, terrestrial and aquatic, seeing that these categories are more or less exclusive of one another; we must accordingly supplement our definition of the term 'aquatic' or 'marine'. For the fact is, some aquatic animals take in water and discharge it again, for the same reason that leads air-breathing animals to inhale air: in other words, with the object of cooling the blood. Others take in water as incidental to their mode of feeding; for as they get their food in the water they cannot but take in water along with their food, and if they take in water they must be provided with some organ for discharging it. Those blooded animals, then, that use water for a purpose analogous to respiration are provided with gills; and such as take in water when catching their prey, with the blow-hole. Similar remarks are applicable to molluscs and crustaceans; for again it is by way of procuring food that these creatures take in water.

Aquatic in different ways, the differences depending on bodily relation to external temperature and on habit of life, are such animals on the one hand as take in air but live in water, and such on the other hand as take in water and are furnished with gills but go upon dry land and get their living there. At present only one animal of the latter kind is known, the so-called cordylus or water-newt; this creature is furnished not with lungs but with gills, but for all that it is a quadruped and fitted for walking on dry land.

In the case of all these animals their nature appears in some kind of a way to have got warped, just as some male animals get to resemble the female, and some female animals the male. The fact is that animals, if they be subjected to a modification in minute organs, are liable to immense modifications in their general configuration. This phenomenon may be observed in the case of gelded animals: only a minute organ of the animal is mutilated, and the creature passes from the male to the female form. We may infer, then, that if in the primary conformation of the embryo an infinitesimally minute but absolutely essential organ sustain a change of magnitude one way or the other, the animal will in one case turn to male and in the other to female; and also that, if the said organ be obliterated altogether, the animal will be of neither one sex nor the other. And so by the occurrence of modification in minute organs it comes to pass that one animal is terrestrial and another aquatic, in both senses of these terms. And, again, some animals are amphibious whilst other animals are not amphibious, owing to the circumstance that in their conformation while in the embryonic condition there got intermixed into them some portion of the matter of which their subsequent food is constituted; for, as was said above, what is in conformity with nature is to every single animal pleasant and agreeable.

Animals then have been categorized into terrestrial and aquatic in three ways, according to their assumption of air or of water, the temperament of their bodies, or the character of their food; and the mode of life of an animal corresponds to the category in which it is found. That is to say, in some cases the animal depends for its terrestrial or aquatic nature on temperament and diet combined, as well as upon its method of respiration; and sometimes on temperament and habits alone.

Of testaceans, some, that are incapable of motion, subsist on fresh water, for, as the sea water dissolves into its constituents, the fresh water from its greater thinness percolates through the grosser parts; in fact, they live on fresh water just as they were originally engendered from the same. Now that fresh water is contained in the sea and can be strained off from it can be proved in a thoroughly practical way. Take a thin vessel of moulded wax, attach a cord to it, and let it down quite empty into the sea: in twenty-four hours it will be found to contain a quantity of water, and the water will be fresh and drinkable.

Sea-anemones feed on such small fishes as come in their way. The mouth of this creature is in the middle of its body; and this fact may be clearly observed in the case of the larger varieties. Like the oyster it has a duct for the outlet of the residuum; and this duct is at the top of the animal. In other words, the sea-anemone corresponds to the inner fleshy part of the oyster, and the stone to which the one creature clings corresponds to the shell which encases the other.

The limpet detaches itself from the rock and goes about in quest of food. Of shell-fish that are mobile, some are carnivorous and live on little fishes, as for instance, the purple murex-and there can be no doubt that the purple murex is carnivorous, as it is caught by a bait of fish; others are carnivorous, but feed also on marine vegetation.

The sea-turtles feed on shell-fish-for, by the way, their mouths are extraordinarily hard; whatever object it seizes, stone or other, it crunches into bits, but when it leaves the water for dry land it browses on grass). These creatures suffer greatly, and oftentimes die when they lie on the surface of the water exposed to a scorching sun; for, when once they have risen to the surface, they find a difficulty in sinking again.

Crustaceans feed in like manner. They are omnivorous; that is to say, they live on stones, slime, sea-weed, and excrement-as for instance the rock-crab-and are also carnivorous. The crawfish or spiny-lobster can get the better of fishes even of the larger species, though in some of them it occasionally finds more than its match. Thus, this animal is so overmastered and cowed by the octopus that it dies of terror if it become aware of an octopus in the same net with itself. The crawfish can master the conger-eel, for owing to the rough spines of the crawfish the eel cannot slip away and elude its hold. The conger-eel, however, devours the octopus, for owing to the slipperiness of its antagonist the octopus can make nothing of it. The crawfish feeds on little fish, capturing them beside its hole or dwelling place; for, by the way, it is found out at sea on rough and stony bottoms, and in such places it makes its den. Whatever it catches, it puts into its mouth with its pincer-like claws, like the common crab. Its nature is to walk straight forward when it has nothing to fear, with its feelers hanging sideways; if it be frightened, it makes its escape backwards, darting off to a great distance. These animals fight one another with their claws, just as rams fight with their horns, raising them and striking their opponents; they are often also seen crowded together in herds. So much for the mode of life of the crustacean.

Molluscs are all carnivorous; and of molluscs the calamary and the sepia are more than a match for fishes even of the large species. The octopus for the most part gathers shellfish, extracts the flesh, and feeds on that; in fact, fishermen recognize their holes by the number of shells lying about. Some say that the octopus devours its own species, but this statement is incorrect; it is doubtless founded on the fact that the creature is often found with its tentacles removed, which tentacles have really been eaten off by the conger.

Fishes, all without exception, feed on spawn in the spawning season; but in other respects the food varies with the varying species. Some fishes are exclusively carnivorous, as the cartilaginous genus, the conger, the channa or Serranus, the tunny, the bass, the synodon or Dentex, the amia, the sea-perch, and the muraena. The red mullet is carnivorous, but feeds also on sea-weed, on shell-fish, and on mud. The grey mullet feeds on mud, the dascyllus on mud and offal, the scarus or parrot-fish and the melanurus on sea-weed, the saupe on offal and sea-weed; the saupe feeds also on zostera, and is the only fish that is captured with a gourd. All fishes devour their own species, with the single exception of the cestreus or mullet; and the conger is especially ravenous in this respect. The cephalus and the mullet in general are the only fish that eat no flesh; this may be inferred from the facts that when caught they are never found with flesh in their intestines, and that the bait used to catch them is not flesh but barley-cake. Every fish of the mullet-kind lives on sea-weed and sand. The cephalus, called by some the 'chelon', keeps near in to the shore, the peraeas keeps out at a distance from it, and feeds on a mucous substance exuding from itself, and consequently is always in a starved condition. The cephalus lives in mud, and is in consequence heavy and slimy; it never feeds on any other fish. As it lives in mud, it has every now and then to make a leap upwards out of the mud so as to wash the slime from off its body. There is no creature known to prey upon the spawn of the cephalus, so that the species is exceedingly numerous; when, however, the is full-grown it is preyed upon by a number of fishes, and especially by the acharnas or bass. Of all fish the mullet is the most voracious and insatiable, and in consequence its belly is kept at full stretch; whenever it is not starving, it may be considered as out of condition. When it is frightened, it hides its head in mud, under the notion that it is hiding its whole body. The synodon is carnivorous and feeds on molluscs. Very often the synodon and the channa cast up their stomachs while chasing smaller fishes; for, be it remembered, fishes have their stomachs close to the mouth, and are not furnished with a gullet.

Some fish then, as has been stated, are carnivorous, and carnivorous only, as the dolphin, the synodon, the gilt-head, the selachians, and the molluscs. Other fishes feed habitually on mud or sea-weed or sea-moss or the so-called stalk-weed or growing plants; as for instance, the phycis, the goby, and the rock-fish; and, by the way, the only meat that the phycis will touch is that of prawns. Very often, however, as has been stated, they devour one another, and especially do the larger ones devour the smaller. The proof of their being carnivorous is the fact that they can be caught with flesh for a bait. The mackerel, the tunny, and the bass are for the most part carnivorous, but they do occasionally feed on sea-weed. The sargue feeds on the leavings of the trigle or red mullet. The red mullet burrows in the mud, when it sets the mud in motion and quits its haunt, the sargue settles down into the place and feeds on what is left behind, and prevents any smaller fish from settling in the immediate vicinity.

Of all fishes the so-called scarus, or parrot, wrasse, is the only one known to chew the cud like a quadruped.

As a general rule the larger fishes catch the smaller ones in their mouths whilst swimming straight after them in the ordinary position; but the selachians, the dolphin, and all the cetacea must first turn over on their backs, as their mouths are placed down below; this allows a fair chance of escape to the smaller fishes, and, indeed, if it were not so, there would be very few of the little fishes left, for the speed and voracity of the dolphin is something marvelous.

Of eels a few here and there feed on mud and on chance morsels of food thrown to them; the greater part of them subsist on fresh water. Eel-breeders are particularly careful to have the water kept perfectly clear, by its perpetually flowing on to flat slabs of stone and then flowing off again; sometimes they coat the eel-tanks with plaster. The fact is that the eel will soon choke if the water is not clear as his gills are peculiarly small. On this account, when fishing for eels, they disturb the water. In the river Strymon eel-fishing takes place at the rising of the Pleiades, because at this period the water is troubled and the mud raised up by contrary winds; unless the water be in this condition, it is as well to leave the eels alone. When dead the eel, unlike the majority of fishes, neither floats on nor rises to the surface; and this is owing to the smallness of the stomach. A few eels are supplied with fat, but the greater part have no fat whatsoever. When removed from the water they can live for five or six days; for a longer period if north winds prevail, for a shorter if south winds. If they are removed in summer from the pools to the tanks they will die; but not so if removed in the winter. They are not capable of holding out against any abrupt change; consequently they often die in large numbers when men engaged in transporting them from one place to another dip them into water particularly cold. They will also die of suffocation if they be kept in a scanty supply of water. This same remark will hold good for fishes in general; for they are suffocated if they be long confined in a short supply of water, with the water kept unchanged-just as animals that respire are suffocated if they be shut up with a scanty supply of air. The eel in some cases lives for seven or eight years. The river-eel feeds on his own species, on grass, or on roots, or on any chance food found in the mud. Their usual feeding-time is at night, and during the day-time they retreat into deep water. And so much for the food of fishes.

Part 3

Of birds, such as have crooked talons are carnivorous without exception, and cannot swallow corn or bread-food even if it be put into their bills in tit-bits; as for instance, the eagle of every variety, the kite, the two species of hawks, to wit, the dove-hawk and the sparrow-hawk-and, by the way, these two hawks differ greatly in size from one another-and the buzzard. The buzzard is of the same size as the kite, and is visible at all seasons of the year. There is also the phene (or lammergeier) and the vulture. The phene is larger than the common eagle and is ashen in colour. Of the vulture there are two varieties: one small and whitish, the other comparatively large and rather more ashen-coloured than white. Further, of birds that fly by night, some have crooked talons, such as the night-raven, the owl, and the eagle-owl. The eagle-owl resembles the common owl in shape, but it is quite as large as the eagle. Again, there is the eleus, the Aegolian owl, and the little horned owl. Of these birds, the eleus is somewhat larger than the barn-door cock, and the Aegolian owl is of about the same size as the eleus, and both these birds hunt the jay; the little horned owl is smaller than the common owl. All these three birds are alike in appearance, and all three are carnivorous.

Again, of birds that have not crooked talons some are carnivorous, such as the swallow. Others feed on grubs, such as the chaffinch, the sparrow, the 'batis', the green linnet, and the titmouse. Of the titmouse there are three varieties. The largest is the finch-titmouse--for it is about the size of a finch; the second has a long tail, and from its habitat is called the hill-titmouse; the third resembles the other two in appearance, but is less in size than either of them. Then come the becca-fico, the black-cap, the bull-finch, the robin, the epilais, the midget-bird, and the golden-crested wren. This wren is little larger than a locust, has a crest of bright red gold, and is in every way a beautiful and graceful little bird. Then the anthus, a bird about the size of a finch; and the mountain-finch, which resembles a finch and is of much the same size, but its neck is blue, and it is named from its habitat; and lastly the wren and the rook. The above-enumerated birds and the like of them feed either wholly or for the most part on grubs, but the following and the like feed on thistles; to wit, the linnet, the thraupis, and the goldfinch. All these birds feed on thistles, but never on grubs or any living thing whatever; they live and roost also on the plants from which they derive their food.

There are other birds whose favourite food consists of insects found beneath the bark of trees; as for instance, the great and the small pie, which are nicknamed the woodpeckers. These two birds resemble one another in plumage and in note, only that the note of the larger bird is the louder of the two; they both frequent the trunks of trees in quest of food. There is also the greenpie, a bird about the size of a turtle-dove, green-coloured all over, that pecks at the bark of trees with extraordinary vigour, lives generally on the branch of a tree, has a loud note, and is mostly found in the Peloponnese. There is another bird called the 'grub-picker' (or tree-creeper), about as small as the penduline titmouse, with speckled plumage of an ashen colour, and with a poor note; it is a variety of the woodpecker.

There are other birds that live on fruit and herbage, such as the wild pigeon or ringdove, the common pigeon, the rock-dove, and the turtle-dove. The ring-dove and the common pigeon are visible at all seasons; the turtledove only in the summer, for in winter it lurks in some hole or other and is never seen. The rock-dove is chiefly visible in the autumn, and is caught at that season; it is larger than the common pigeon but smaller than the wild one; it is generally caught while drinking. These pigeons bring their young ones with them when they visit this country. All our other birds come to us in the early summer and build their nests here, and the greater part of them rear their young on animal food, with the sole exception of the pigeon and its varieties.

The whole genus of birds may be pretty well divided into such as procure their food on dry land, such as frequent rivers and lakes, and such as live on or by the sea.

Of water-birds such as are web-footed live actually on the water, while such as are split-footed live by the edge of it-and, by the way, water-birds that are not carnivorous live on water-plants, (but most of them live on fish), like the heron and the spoonbill that frequent the banks of lakes and rivers; and the spoonbill, by the way, is less than the common heron, and has a long flat bill. There are furthermore the stork and the seamew; and the seamew, by the way, is ashen-coloured. There is also the schoenilus, the cinclus, and the white-rump. Of these smaller birds the last mentioned is the largest, being about the size of the common thrush; all three may be described as 'wag-tails'. Then there is the scalidris, with plumage ashen-grey, but speckled. Moreover, the family of the halcyons or kingfishers live by the waterside. Of kingfishers there are two varieties; one that sits on reeds and sings; the other, the larger of the two, is without a note. Both these varieties are blue on the back. There is also the trochilus (or sandpiper). The halcyon also, including a variety termed the cerylus, is found near the seaside. The crow also feeds on such animal life as is cast up on the beach, for the bird is omnivorous. There are also the white gull, the cepphus, the aethyia, and the charadrius.

Of web-footed birds, the larger species live on the banks of rivers and lakes; as the swan, the duck, the coot, the grebe, and the teal-a bird resembling the duck but less in size-and the water-raven or cormorant. This bird is the size of a stork, only that its legs are shorter; it is web-footed and is a good swimmer; its plumage is black. It roosts on trees, and is the only one of all such birds as these that is found to build its nest in a tree. Further there is the large goose, the little gregarious goose, the vulpanser, the horned grebe, and the penelops. The sea-eagle lives in the neighbourhood of the sea and seeks its quarry in lagoons.

A great number of birds are omnivorous. Birds of prey feed on any animal or bird, other than a bird of prey, that they may catch. These birds never touch one of their own genus, whereas fishes often devour members actually of their own species.

Birds, as a rule, are very spare drinkers. In fact birds of prey never drink at all, excepting a very few, and these drink very rarely; and this last observation is peculiarly applicable to the kestrel. The kite has been seen to drink, but he certainly drinks very seldom.

Part 4

Animals that are coated with tessellates-such as the lizard and the other quadrupeds, and the serpents-are omnivorous: at all events they are carnivorous and graminivorous; and serpents, by the way, are of all animals the greatest gluttons.

Tessellated animals are spare drinkers, as are also all such animals as have a spongy lung, and such a lung, scantily supplied with blood, is found in all oviparous animals. Serpents, by the by, have an insatiate appetite for wine; consequently, at times men hunt for snakes by pouring wine into saucers and putting them into the interstices of walls, and the creatures are caught when inebriated. Serpents are carnivorous, and whenever they catch an animal they extract all its juices and eject the creature whole. And, by the way, this is done by all other creatures of similar habits, as for instance the spider; only that the spider sucks out the juices of its prey outside, and the serpent does so in its belly. The serpent takes any food presented to him, eats birds and animals, and swallows eggs entire. But after taking his prey he stretches himself until he stands straight out to the very tip, and then he contracts and squeezes himself into little compass, so that the swallowed mass may pass down his outstretched body; and this action on his part is due to the tenuity and length of his gullet. Spiders and snakes can both go without food for a long time; and this remark may be verified by observation of specimens kept alive in the shops of the apothecaries.

Part 5

Of viviparous quadrupeds such as are fierce and jag-toothed are without exception carnivorous; though, by the way, it is stated of the wolf, but of no other animal, that in extremity of hunger it will eat a certain kind of earth. These carnivorous animals never eat grass except when they are sick, just as dogs bring on a vomit by eating grass and thereby purge themselves.

The solitary wolf is more apt to attack man than the wolf that goes with a pack.

The animal called 'glanus' by some and 'hyena' by others is as large as a wolf, with a mane like a horse, only that the hair is stiffer and longer and extends over the entire length of the chine. It will lie in wait for a man and chase him, and will inveigle a dog within its reach by making a noise that resembles the retching noise of a man vomiting. It is exceedingly fond of putrefied flesh, and will burrow in a graveyard to gratify this propensity.

The bear is omnivorous. It eats fruit, and is enabled by the suppleness of its body to climb a tree; it also eats vegetables, and it will break up a hive to get at the honey; it eats crabs and ants also, and is in a general way carnivorous. It is so powerful that it will attack not only the deer but the wild boar, if it can take it unawares, and also the bull. After coming to close quarters with the bull it falls on its back in front of the animal, and, when the bull proceeds to butt, the bear seizes hold of the bull's horns with its front paws, fastens its teeth into his shoulder, and drags him down to the ground. For a short time together it can walk erect on its hind legs. All the flesh it eats it first allows to become carrion.

The lion, like all other savage and jag-toothed animals, is carnivorous. It devours its food greedily and fiercely, and often swallows its prey entire without rending it at all; it will then go fasting for two or three days together, being rendered capable of this abstinence by its previous surfeit. It is a spare drinker. It discharges the solid residuum in small quantities, about every other day or at irregular intervals, and the substance of it is hard and dry like the excrement of a dog. The wind discharged from off its stomach is pungent, and its urine emits a strong odour, a phenomenon which, in the case of dogs, accounts for their habit of sniffing at trees; for, by the way, the lion, like the dog, lifts its leg to void its urine. It infects the food it eats with a strong smell by breathing on it, and when the animal is cut open an overpowering vapour exhales from its inside.

Some wild quadrupeds feed in lakes and rivers; the seal is the only one that gets its living on the sea. To the former class of animals belong the so-called castor, the satyrium, the otter, and the so-called latax, or beaver. The beaver is flatter than the otter and has strong teeth; it often at night-time emerges from the water and goes nibbling at the bark of the aspens that fringe the riversides. The otter will bite a man, and it is said that whenever it bites it will never let go until it hears a bone crack. The hair of the beaver is rough, intermediate in appearance between the hair of the seal and the hair of the deer.

Part 6

Jag-toothed animals drink by lapping, as do also some animals with teeth differently formed, as the mouse. Animals whose upper and lower teeth meet evenly drink by suction, as the horse and the ox; the bear neither laps nor sucks, but gulps down his drink. Birds, a rule, drink by suction, but the long necked birds stop and elevate their heads at intervals; the purple coot is the only one (of the long-necked birds) that swallows water by gulps.

Horned animals, domesticated or wild, and all such as are not jag-toothed, are all frugivorous and graminivorous, save under great stress of hunger. The pig is an exception, it cares little for grass or fruit, but of all animals it is the fondest of roots, owing to the fact that its snout is peculiarly adapted for digging them out of the ground; it is also of all animals the most easily pleased in the matter of food. It takes on fat more rapidly in proportion to its size than any other animal; in fact, a pig can be fattened for the market in sixty days. Pig-dealers can tell the amount of flesh taken on, by having first weighed the animal while it was being starved. Before the fattening process begins, the creature must be starved for three days; and, by the way, animals in general will take on fat if subjected previously to a course of starvation; after the three days of starvation, pig-breeders feed the animal lavishly. Breeders in Thrace, when fattening pigs, give them a drink on the first day; then they miss one, and then two days, then three and four, until the interval extends over seven days. The pigs' meat used for fattening is composed of barley, millet, figs, acorns, wild pears, and cucumbers. These animals-and other animals that have warm bellies-are fattened by repose. (Pigs also fatten the better by being allowed to wallow in mud. They like to feed in batches of the same age. A pig will give battle even to a wolf.) If a pig be weighed when living, you may calculate that after death its flesh will weigh five-sixths of that weight, and the hair, the blood, and the rest will weigh the other sixth. When suckling their young, swinelike all other animals-get attenuated. So much for these animals.

Part 7

Cattle feed on corn and grass, and fatten on vegetables that tend to cause flatulency, such as bitter vetch or bruised beans or bean-stalks. The older ones also will fatten if they be fed up after an incision has been made into their hide, and air blown thereinto. Cattle will fatten also on barley in its natural state or on barley finely winnowed, or on sweet food, such as figs, or pulp from the wine-press, or on elm-leaves. But nothing is so fattening as the heat of the sun and wallowing in warm waters. If the horns of young cattle be smeared with hot wax, you may mold them to any shape you please, and cattle are less subject to disease of the hoof if you smear the horny parts with wax, pitch, or olive oil. Herded cattle suffer more when they are forced to change their pasture ground by frost than when snow is the cause of change. Cattle grow all the more in size when they are kept from sexual commerce over a number of years; and it is with a view to growth in size that in Epirus the so-called Pyrrhic kine are not allowed intercourse with the bull until they are nine years old; from which circumstance they are nicknamed the 'unbulled' kine. Of these Pyrrhic cattle, by the way, they say that there are only about four hundred in the world, that they are the private property of the Epirote royal family, that they cannot thrive out of Epirus, and that people elsewhere have tried to rear them, but without success.

Part 8

Horses, mules, and asses feed on corn and grass, but are fattened chiefly by drink. Just in proportion as beasts of burden drink water, so will they more or less enjoy their food, and a place will give good or bad feeding according as the water is good or bad. Green corn, while ripening, will give a smooth coat; but such corn is injurious if the spikes are too stiff and sharp. The first crop of clover is unwholesome, and so is clover over which ill-scented water runs; for the clover is sure to get the taint of the water. Cattle like clear water for drinking; but the horse in this respect resembles the camel, for the camel likes turbid and thick water, and will never drink from a stream until he has trampled it into a turbid condition. And, by the way, the camel can go without water for as much as four days, but after that when he drinks, he drinks in immense quantities.

Part 9

The elephant at the most can eat nine Macedonian medimni of fodder at one meal; but so large an amount is unwholesome. As a general rule it can take six or seven medimni of fodder, five medimni of wheat, and five mareis of wine-six cotylae going to the maris. An elephant has been known to drink right off fourteen Macedonian metretae of water, and another metretae later in the day.

Camels live for about thirty years; in some exceptional cases they live much longer, and instances have been known of their living to the age of a hundred. The elephant is said by some to live for about two hundred years; by others, for three hundred.

Part 10

Sheep and goats are graminivorous, but sheep browse assiduously and steadily, whereas goats shift their ground rapidly, and browse only on the tips of the herbage. Sheep are much improved in condition by drinking, and accordingly they give the flocks salt every five days in summer, to the extent of one medimnus to the hundred sheep, and this is found to render a flock healthier and fatter. In fact they mix salt with the greater part of their food; a large amount of salt is mixed into their bran (for the reason that they drink more when thirsty), and in autumn they get cucumbers with a sprinkling of salt on them; this admixture of salt in their food tends also to increase the quantity of milk in the ewes. If sheep be kept on the move at midday they will drink more copiously towards evening; and if the ewes be fed with salted food as the lambing season draws near they will get larger udders. Sheep are fattened by twigs of the olive or of the oleaster, by vetch, and bran of every kind; and these articles of food fatten all the more if they be first sprinkled with brine. Sheep will take on flesh all the better if they be first put for three days through a process of starving. In autumn, water from the north is more wholesome for sheep than water from the south. Pasture grounds are all the better if they have a westerly aspect.

Sheep will lose flesh if they be kept overmuch on the move or be subjected to any hardship. In winter time shepherds can easily distinguish the vigorous sheep from the weakly, from the fact that the vigorous sheep are covered with hoar-frost while the weakly ones are quite free of it; the fact being that the weakly ones feeling oppressed with the burden shake themselves and so get rid of it. The flesh of all quadrupeds deteriorates in marshy pastures, and is the better on high grounds. Sheep that have flat tails can stand the winter better than long-tailed sheep, and short-fleeced sheep than the shaggy-fleeced; and sheep with crisp wool stand the rigour of winter very poorly. Sheep are healthier than goats, but goats are stronger than sheep. (The fleeces and the wool of sheep that have been killed by wolves, as also the clothes made from them, are exceptionally infested with lice.)

Part 11

Of insects, such as have teeth are omnivorous; such as have a tongue feed on liquids only, extracting with that organ juices from all quarters. And of these latter some may be called omnivorous, inasmuch as they feed on every kind of juice, as for instance, the common fly; others are blood-suckers, such as the gadfly and the horse-fly, others again live on the juices of fruits and plants. The bee is the only insect that invariably eschews whatever is rotten; it will touch no article of food unless it have a sweet-tasting juice, and it is particularly fond of drinking water if it be found bubbling up clear from a spring underground.

So much for the food of animals of the leading genera.

Part 12

The habits of animals are all connected with either breeding and the rearing of young, or with the procuring a due supply of food; and these habits are modified so as to suit cold and heat and the variations of the seasons. For all animals have an instinctive perception of the changes of temperature, and, just as men seek shelter in houses in winter, or as men of great possessions spend their summer in cool places and their winter in sunny ones, so also all animals that can do so shift their habitat at various seasons.

Some creatures can make provision against change without stirring from their ordinary haunts; others migrate, quitting Pontus and the cold countries after the autumnal equinox to avoid the approaching winter, and after the spring equinox migrating from warm lands to cool lands to avoid the coming heat. In some cases they migrate from places near at hand, in others they may be said to come from the ends of the world, as in the case of the crane; for these birds migrate from the steppes of Scythia to the marshlands south of Egypt where the Nile has its source. And it is here, by the way, that they are said to fight with the pygmies; and the story is not fabulous, but there is in reality a race of dwarfish men, and the horses are little in proportion, and the men live in caves underground. Pelicans also migrate, and fly from the Strymon to the Ister, and breed on the banks of this river. They depart in flocks, and the birds in front wait for those in the rear, owing to the fact that when the flock is passing over the intervening mountain range, the birds in the rear lose sight of their companions in the van.

Fishes also in a similar manner shift their habitat now out of the Euxine and now into it. In winter they move from the outer sea in towards land in quest of heat; in summer they shift from shallow waters to the deep sea to escape the heat.

Weakly birds in winter and in frosty weather come down to the plains for warmth, and in summer migrate to the hills for coolness. The more weakly an animal is the greater hurry will it be in to migrate on account of extremes of temperature, either hot or cold; thus the mackerel migrates in advance of the tunnies, and the quail in advance of the cranes. The former migrates in the month of Boedromion, and the latter in the month of Maemacterion. All creatures are fatter in migrating from cold to heat than in migrating from heat to cold; thus the quail is fatter when he emigrates in autumn than when he arrives in spring. The migration from cold countries is contemporaneous with the close of the hot season. Animals are in better trim for breeding purposes in spring-time, when they change from hot to cool lands.

Of birds, the crane, as has been said, migrates from one end of the world to the other; they fly against the wind. The story told about the stone is untrue: to wit, that the bird, so the story goes, carries in its inside a stone by way of ballast, and that the stone when vomited up is a touchstone for gold.

The cushat and the rock-dove migrate, and never winter in our country, as is the case also with the turtle-dove; the common pigeon, however, stays behind. The quail also migrates; only, by the way, a few quails and turtle-doves may stay behind here and there in sunny districts. Cushats and turtle-doves flock together, both when they arrive and when the season for migration comes round again. When quails come to land, if it be fair weather or if a north wind is blowing, they will pair off and manage pretty comfortably; but if a southerly wind prevail they are greatly distressed owing to the difficulties in the way of flight, for a southerly wind is wet and violent. For this reason bird-catchers are never on the alert for these birds during fine weather, but only during the prevalence of southerly winds, when the bird from the violence of the wind is unable to fly. And, by the way, it is owing to the distress occasioned by the bulkiness of its body that the bird always screams while flying: for the labour is severe. When the quails come from abroad they have no leaders, but when they migrate hence, the glottis flits along with them, as does also the landrail, and the eared owl, and the corncrake. The corncrake calls them in the night, and when the birdcatchers hear the croak of the bird in the nighttime they know that the quails are on the move. The landrail is like a marsh bird, and the glottis has a tongue that can project far out of its beak. The eared owl is like an ordinary owl, only that it has feathers about its ears; by some it is called the night-raven. It is a great rogue of a bird, and is a capital mimic; a bird-catcher will dance before it and, while the bird is mimicking his gestures, the accomplice comes behind and catches it. The common owl is caught by a similar trick.

As a general rule all birds with crooked talons are short-necked, flat-tongued, and disposed to mimicry. The Indian bird, the parrot, which is said to have a man's tongue, answers to this description; and, by the way, after drinking wine, the parrot becomes more saucy than ever.

Of birds, the following are migratory-the crane, the swan, the pelican, and the lesser goose.

Part 13

Of fishes, some, as has been observed, migrate from the outer seas in towards shore, and from the shore towards the outer seas, to avoid the extremes of cold and heat.

Fish living near to the shore are better eating than deep-sea fish. The fact is they have more abundant and better feeding, for wherever the sun's heat can reach vegetation is more abundant, better in quality, and more delicate, as is seen in any ordinary garden. Further, the black shore-weed grows near to shore; the other shore-weed is like wild weed. Besides, the parts of the sea near to shore are subjected to a more equable temperature; and consequently the flesh of shallow-water fishes is firm and consistent, whereas the flesh of deep-water fishes is flaccid and watery.

The following fishes are found near into the shore-the synodon, the black bream, the merou, the gilthead, the mullet, the red mullet, the wrasse, the weaver, the callionymus, the goby, and rock-fishes of all kinds. The following are deep-sea fishes--the trygon, the cartilaginous fishes, the white conger, the serranus, the erythrinus, and the glaucus. The braize, the sea-scorpion, the black conger, the muraena, and the piper or sea-cuckoo are found alike in shallow and deep waters. These fishes, however, vary for various localities; for instance, the goby and all rock-fish are fat off the coast of Crete. Again, the tunny is out of season in summer, when it is being preyed on by its own peculiar louse-parasite, but after the rising of Arcturus, when the parasite has left it, it comes into season again. A number of fish also are found in sea-estuaries; such as the saupe, the gilthead, the red mullet, and, in point of fact, the greater part of the gregarious fishes. The bonito also is found in such waters, as, for instance, off the coast of Alopeconnesus; and most species of fishes are found in Lake Bistonis. The coly-mackerel as a rule does not enter the Euxine, but passes the summer in the Propontis, where it spawns, and winters in the Aegean. The tunny proper, the pelamys, and the bonito penetrate into the Euxine in summer and pass the summer there; as do also the greater part of such fish as swim in shoals with the currents, or congregate in shoals together. And most fish congregate in shoals, and shoal-fishes in all cases have leaders.

Fish penetrate into the Euxine for two reasons, and firstly for food. For the feeding is more abundant and better in quality owing to the amount of fresh river-water that discharges into the sea, and moreover, the large fishes of this inland sea are smaller than the large fishes of the outer sea. In point of fact, there is no large fish in the Euxine excepting the dolphin and the porpoise, and the dolphin is a small variety; but as soon as you get into the outer sea the big fishes are on the big scale. Furthermore, fish penetrate into this sea for the purpose of breeding; for there are recesses there favourable for spawning, and the fresh and exceptionally sweet water has an invigorating effect upon the spawn. After spawning, when the young fishes have attained some size, the parent fish swim out of the Euxine immediately after the rising of the Pleiades. If winter comes in with a southerly wind, they swim out with more or less of deliberation; but, if a north wind be blowing, they swim out with greater rapidity, from the fact that the breeze is favourable to their own course. And, by the way, the young fish are caught about this time in the neighbourhood of Byzantium very small in size, as might have been expected from the shortness of their sojourn in the Euxine. The shoals in general are visible both as they quit and enter the Euxine. The trichiae, however, only can be caught during their entry, but are never visible during their exit; in point of fact, when a trichia is caught running outwards in the neighbourhood of Byzantium, the fishermen are particularly careful to cleanse their nets, as the circumstance is so singular and exceptional. The way of accounting for this phenomenon is that this fish, and this one only, swims northwards into the Danube, and then at the point of its bifurcation swims down southwards into the Adriatic. And, as a proof that this theory is correct, the very opposite phenomenon presents itself in the Adriatic; that is to say, they are not caught in that sea during their entry, but are caught during their exit.

Tunny-fish swim into the Euxine keeping the shore on their right, and swim out of it with the shore upon their left. It is stated that they do so as being naturally weak-sighted, and seeing better with the right eye.

During the daytime shoal-fish continue on their way, but during the night they rest and feed. But if there be moonlight, they continue their journey without resting at all. Some people accustomed to sea-life assert that shoal-fish at the period of the winter solstice never move at all, but keep perfectly still wherever they may happen to have been overtaken by the solstice, and this lasts until the equinox.

The coly-mackerel is caught more frequently on entering than on quitting the Euxine. And in the Propontis the fish is at its best before the spawning season. Shoal-fish, as a rule, are caught in greater quantities as they leave the Euxine, and at that season they are in the best condition. At the time of their entrance they are caught in very plump condition close to shore, but those are in comparatively poor condition that are caught farther out to sea. Very often, when the coly-mackerel and the mackerel are met by a south wind in their exit, there are better catches to the southward than in the neighbourhood of Byzantium. So much then for the phenomenon of migration of fishes.

Now the same phenomenon is observed in fishes as in terrestrial animals in regard to hibernation: in other words, during winter fishes take to concealing themselves in out of the way places, and quit their places of concealment in the warmer season. But, by the way, animals go into concealment by way of refuge against extreme heat, as well as against extreme cold. Sometimes an entire genus will thus seek concealment; in other cases some species will do so and others will not. For instance, the shell-fish seek concealment without exception, as is seen in the case of those dwelling in the sea, the purple murex, the ceryx, and all such like; but though in the case of the detached species the phenomenon is obvious-for they hide themselves, as is seen in the scallop, or they are provided with an operculum on the free surface, as in the case of land snails-in the case of the non-detached the concealment is not so clearly observed. They do not go into hiding at one and the same season; but the snails go in winter, the purple murex and the ceryx for about thirty days at the rising of the Dog-star, and the scallop at about the same period. But for the most part they go into concealment when the weather is either extremely cold or extremely hot.

Part 14

Insects almost all go into hiding, with the exception of such of them as live in human habitations or perish before the completion of the year. They hide in the winter; some of them for several days, others for only the coldest days, as the bee. For the bee also goes into hiding: and the proof that it does so is that during a certain period bees never touch the food set before them, and if a bee creeps out of the hive, it is quite transparent, with nothing whatsoever in its stomach; and the period of its rest and hiding lasts from the setting of the Pleiades until springtime.

Animals take their winter-sleep or summer-sleep by concealing themselves in warm places, or in places where they have been used to lie concealed.

Part 15

Several blooded animals take this sleep, such as the pholidotes or tessellates, namely, the serpent, the lizard, the gecko, and the river. crocodile, all of which go into hiding for four months in the depth of winter, and during that time eat nothing. Serpents in general burrow underground for this purpose; the viper conceals itself under a stone.

A great number of fishes also take this sleep, and notably, the hippurus and coracinus in winter time; for, whereas fish in general may be caught at all periods of the year more or less, there is this singularity observed in these fishes, that they are caught within a certain fixed period of the year, and never by any chance out of it. The muraena also hides, and the orphus or sea-perch, and the conger. Rock-fish pair off, male and female, for hiding (just as for breeding); as is observed in the case of the species of wrasse called the thrush and the owzel, and in the perch.

The tunny also takes a sleep in winter in deep waters, and gets exceedingly fat after the sleep. The fishing season for the tunny begins at the rising of the Pleiades and lasts, at the longest, down to the setting of Arcturus; during the rest of the year they are hid and enjoying immunity. About the time of hibernation a few tunnies or other hibernating fishes are caught while swimming about, in particularly warm localities and in exceptionally fine weather, or on nights of full moon; for the fishes are induced (by the warmth or the light) to emerge for a while from their lair in quest of food.

Most fish are at their best for the table during the summer or winter sleep.

The primas-tunny conceals itself in the mud; this may be inferred from the fact that during a particular period the fish is never caught, and that, when it is caught after that period, it is covered with mud and has its fins damaged. In the spring these tunnies get in motion and proceed towards the coast, coupling and breeding, and the females are now caught full of spawn. At this time they are considered as in season, but in autumn and in winter as of inferior quality; at this time also the males are full of milt. When the spawn is small, the fish is hard to catch, but it is easily caught when the spawn gets large, as the fish is then infested by its parasite. Some fish burrow for sleep in the sand and some in mud, just keeping their mouths outside.

Most fishes hide, then, during the winter only, but crustaceans, the rock-fish, the ray, and the cartilaginous species hide only during extremely severe weather, and this may be inferred from the fact that these fishes are never by any chance caught when the weather is extremely cold. Some fishes, however, hide during the summer, as the glaucus or grey-back; this fish hides in summer for about sixty days. The hake also and the gilthead hide; and we infer that the hake hides over a lengthened period from the fact that it is only caught at long intervals. We are led also to infer that fishes hide in summer from the circumstance that the takes of certain fish are made between the rise and setting of certain constellations: of the Dog-star in particular, the sea at this period being upturned from the lower depths. This phenomenon may be observed to best advantage in the Bosporus; for the mud is there brought up to the surface and the fish are brought up along with it. They say also that very often, when the sea-bottom is dredged, more fish will be caught by the second haul than by the first one. Furthermore, after very heavy rains numerous specimens become visible of creatures that at other times are never seen at all or seen only at intervals.

Part 16

A great number of birds also go into hiding; they do not all migrate, as is generally supposed, to warmer countries. Thus, certain birds (as the kite and the swallow) when they are not far off from places of this kind, in which they have their permanent abode, betake themselves thither; others, that are at a distance from such places, decline the trouble of migration and simply hide themselves where they are. Swallows, for instance, have been often found in holes, quite denuded of their feathers, and the kite on its first emergence from torpidity has been seen to fly from out some such hiding-place. And with regard to this phenomenon of periodic torpor there is no distinction observed, whether the talons of a bird be crooked or straight; for instance, the stork, the owzel, the turtle-dove, and the lark, all go into hiding. The case of the turtledove is the most notorious of all, for we would defy anyone to assert that he had anywhere seen a turtle-dove in winter-time; at the beginning of the hiding time it is exceedingly plump, and during this period it moults, but retains its plumpness. Some cushats hide; others, instead of hiding, migrate at the same time as the swallow. The thrush and the starling hide; and of birds with crooked talons the kite and the owl hide for a few days.

Part 17

Of viviparous quadrupeds the porcupine and the bear retire into concealment. The fact that the bear hides is well established, but there are doubts as to its motive for so doing, whether it be by reason of the cold or from some other cause. About this period the male and the female become so fat as to be hardly capable of motion. The female brings forth her young at this time, and remains in concealment until it is time to bring the cubs out; and she brings them out in spring, about three months after the winter solstice. The bear hides for at least forty days; during fourteen of these days it is said not to move at all, but during most of the subsequent days it moves, and from time to time wakes up. A she-bear in pregnancy has either never been caught at all or has been caught very seldom. There can be no doubt but that during this period they eat nothing; for in the first place they never emerge from their hiding-place, and further, when they are caught, their belly and intestines are found to be quite empty. It is also said that from no food being taken the gut almost closes up, and that in consequence the animal on first emerging takes to eating arum with the view of opening up and distending the gut.

The dormouse actually hides in a tree, and gets very fat at that period; as does also the white mouse of Pontus.

(Of animals that hide or go torpid some slough off what is called their 'old-age'. This name is applied to the outermost skin, and to the casing that envelops the developing organism.)

In discussing the case of terrestrial vivipara we stated that the reason for the bear's seeking concealment is an open question. We now proceed to treat of the tessellates. The tessellates for the most part go into hiding, and if their skin is soft they slough off their 'old-age', but not if the skin is shell-like, as is the shell of the tortoise-for, by the way, the tortoise and the fresh water tortoise belong to the tessellates. Thus, the old-age is sloughed off by the gecko, the lizard, and above all, by serpents; and they slough off the skin in springtime when emerging from their torpor, and again in the autumn. Vipers also slough off their skin both in spring and in autumn, and it is not the case, as some aver, that this species of the serpent family is exceptional in not sloughing. When the serpent begins to slough, the skin peels off at first from the eyes, so that any one ignorant of the phenomenon would suppose the animal were going blind; after that it peels off the head, and so on, until the creature presents to view only a white surface all over. The sloughing goes on for a day and a night, beginning with the head and ending with the tail. During the sloughing of the skin an inner layer comes to the surface, for the creature emerges just as the embryo from its afterbirth.

All insects that slough at all slough in the same way; as the silphe, and the empis or midge, and all the coleoptera, as for instance the cantharus-beetle. They all slough after the period of development; for just as the afterbirth breaks from off the young of the vivipara so the outer husk breaks off from around the young of the vermipara, in the same way both with the bee and the grasshopper. The cicada the moment after issuing from the husk goes and sits upon an olive tree or a reed; after the breaking up of the husk the creature issues out, leaving a little moisture behind, and after a short interval flies up into the air and sets a. chirping.

Of marine animals the crawfish and the lobster slough sometimes in the spring, and sometimes in autumn after parturition. Lobsters have been caught occasionally with the parts about the thorax soft, from the shell having there peeled off, and the lower parts hard, from the shell having not yet peeled off there; for, by the way, they do not slough in the same manner as the serpent. The crawfish hides for about five months. Crabs also slough off their old-age; this is generally allowed with regard to the soft-shelled crabs, and it is said to be the case with the testaceous kind, as for instance with the large 'granny' crab. When these animals slough their shell becomes soft all over, and as for the crab, it can scarcely crawl. These animals also do not cast their skins once and for all, but over and over again.

So much for the animals that go into hiding or torpidity, for the times at which, and the ways in which, they go; and so much also for the animals that slough off their old-age, and for the times at which they undergo the process.

Part 18

Animals do not all thrive at the same seasons, nor do they thrive alike during all extremes of weather. Further animals of diverse species are in a diverse way healthy or sickly at certain seasons; and, in point of fact, some animals have ailments that are unknown to others. Birds thrive in times of drought, both in their general health and in regard to parturition, and this is especially the case with the cushat; fishes, however, with a few exceptions, thrive best in rainy weather; on the contrary rainy seasons are bad for birds-and so by the way is much drinking-and drought is bad for fishes. Birds of prey, as has been already stated, may in a general way be said never to drink at all, though Hesiod appears to have been ignorant of the fact, for in his story about the siege of Ninus he represents the eagle that presided over the auguries as in the act of drinking; all other birds drink, but drink sparingly, as is the case also with all other spongy-lunged oviparous animals. Sickness in birds may be diagnosed from their plumage, which is ruffled when they are sickly instead of lying smooth as when they are well.

Part 19

The majority of fishes, as has been stated, thrive best in rainy seasons. Not only have they food in greater abundance at this time, but in a general way rain is wholesome for them just as it is for vegetation-for, by the way, kitchen vegetables, though artificially watered, derive benefit from rain; and the same remark applies even to reeds that grow in marshes, as they hardly grow at all without a rainfall. That rain is good for fishes may be inferred from the fact that most fishes migrate to the Euxine for the summer; for owing to the number of the rivers that discharge into this sea its water is exceptionally fresh, and the rivers bring down a large supply of food. Besides, a great number of fishes, such as the bonito and the mullet, swim up the rivers and thrive in the rivers and marshes. The sea-gudgeon also fattens in the rivers, and, as a rule, countries abounding in lagoons furnish unusually excellent fish. While most fishes, then, are benefited by rain, they are chiefly benefited by summer rain; or we may state the case thus, that rain is good for fishes in spring, summer, and autumn, and fine dry weather in winter. As a general rule what is good for men is good for fishes also.

Fish do not thrive in cold places, and those fish suffer most in severe winters that have a stone in their head, as the chromis, the bass, the sciaena, and the braize; for owing to the stone they get frozen with the cold, and are thrown up on shore.

Whilst rain is wholesome for most fishes, it is, on the contrary, unwholesome for the mullet, the cephalus, and the so-called marinus, for rain superinduces blindness in most of these fishes, and all the more rapidly if the rainfall be superabundant. The cephalus is peculiarly subject to this malady in severe winters; their eyes grow white, and when caught they are in poor condition, and eventually the disease kills them. It would appear that this disease is due to extreme cold even more than to an excessive rainfall; for instance, in many places and more especially in shallows off the coast of Nauplia, in the Argolid, a number of fishes have been known to be caught out at sea in seasons of severe cold. The gilthead also suffers in winter; the acharnas suffers in summer, and loses condition. The coracine is exceptional among fishes in deriving benefit from drought, and this is due to the fact that heat and drought are apt to come together.

Particular places suit particular fish; some are naturally fish of the shore, and some of the deep sea, and some are at home in one or the other of these regions, and others are common to the two and are at home in both. Some fish will thrive in one particular spot, and in that spot only. As a general rule it may be said that places abounding in weeds are wholesome; at all events, fishes caught in such places are exceptionally fat: that is, such fishes a a habit all sorts of localities as well. The fact is that weed-eating fish find abundance of their special food in such localities, and carnivorous fish find an unusually large number of smaller fish. It matters also whether the wind be from the north or south: the longer fish thrive better when a north wind prevails, and in summer at one and the same spot more long fish will be caught than flat fish with a north wind blowing.

The tunny and the sword-fish are infested with a parasite about the rising of the Dog-star; that is to say, about this time both these fishes have a grub beside their fins that is nicknamed the 'gadfly'. It resembles the scorpion in shape, and is about the size of the spider. So acute is the pain it inflicts that the sword-fish will often leap as high out of the water as a dolphin; in fact, it sometimes leaps over the bulwarks of a vessel and falls back on the deck. The tunny delights more than any other fish in the heat of the sun. It will burrow for warmth in the sand in shallow waters near to shore, or will, because it is warm, disport itself on the surface of the sea.

The fry of little fish escape by being overlooked, for it is only the larger ones of the small species that fishes of the large species will pursue. The greater part of the spawn and the fry of fishes is destroyed by the heat of the sun, for whatever of them the sun reaches it spoils.

Fish are caught in greatest abundance before sunrise and after sunset, or, speaking generally, just about sunset and sunrise. Fishermen haul up their nets at these times, and speak of the hauls then made as the 'nick-of-time' hauls. The fact is, that at these times fishes are particularly weak-sighted; at night they are at rest, and as the light grows stronger they see comparatively well.

We know of no pestilential malady attacking fishes, such as those which attack man, and horses and oxen among the quadrupedal vivipara, and certain species of other genera, domesticated and wild; but fishes do seem to suffer from sickness; and fishermen infer this from the fact that at times fishes in poor condition, and looking as though they were sick, and of altered colour, are caught in a large haul of well-conditioned fish of their own species. So much for sea-fishes.

Part 20

River-fish and lake-fish also are exempt from diseases of a pestilential character, but certain species are subject to special and peculiar maladies. For instance, the sheat-fish just before the rising of the Dog-star, owing to its swimming near the surface of the water, is liable to sunstroke, and is paralysed by a loud peal of thunder. The carp is subject to the same eventualities but in a lesser degree. The sheatfish is destroyed in great quantities in shallow waters by the serpent called the dragon. In the balerus and tilon a worm is engendered about the rising of the Dog-star, that sickens these fish and causes them to rise towards the surface, where they are killed by the excessive heat. The chalcis is subject to a very violent malady; lice are engendered underneath their gills in great numbers, and cause destruction among them; but no other species of fish is subject to any such malady.

If mullein be introduced into water it will kill fish in its vicinity. It is used extensively for catching fish in rivers and ponds; by the Phoenicians it is made use of also in the sea.

There are two other methods employed for catch-fish. It is a known fact that in winter fishes emerge from the deep parts of rivers and, by the way, at all seasons fresh water is tolerably cold. A trench accordingly is dug leading into a river, and wattled at the river end with reeds and stones, an aperture being left in the wattling through which the river water flows into the trench; when the frost comes on the fish can be taken out of the trench in weels. Another method is adopted in summer and winter alike. They run across a stream a dam composed of brushwood and stones leaving a small open space, and in this space they insert a weel; they then coop the fish in towards this place, and draw them up in the weel as they swim through the open space.

Shell-fish, as a rule, are benefited by rainy weather. The purple murex is an exception; if it be placed on a shore near to where a river discharges, it will die within a day after tasting the fresh water. The murex lives for about fifty days after capture; during this period they feed off one another, as there grows on the shell a kind of sea-weed or sea-moss; if any food is thrown to them during this period, it is said to be done not to keep them alive, but to make them weigh more.

To shell-fish in general drought is unwholesome. During dry weather they decrease in size and degenerate in quality; and it is during such weather that the red scallop is found in more than usual abundance. In the Pyrrhaean Strait the clam was exterminated, partly by the dredging-machine used in their capture, and partly by long-continued droughts. Rainy weather is wholesome to the generality of shellfish owing to the fact that the sea-water then becomes exceptionally sweet. In the Euxine, owing to the coldness of the climate, shellfish are not found: nor yet in rivers, excepting a few bivalves here and there. Univalves, by the way, are very apt to freeze to death in extremely cold weather. So much for animals that live in water.

Part 21

To turn to quadrupeds, the pig suffers from three diseases, one of which is called branchos, a disease attended with swellings about the windpipe and the jaws. It may break out in any part of the body; very often it attacks the foot, and occasionally the ear; the neighbouring parts also soon rot, and the decay goes on until it reaches the lungs, when the animal succumbs. The disease develops with great rapidity, and the moment it sets in the animal gives up eating. The swineherds know but one way to cure it, namely, by complete excision, when they detect the first signs of the disease. There are two other diseases, which are both alike termed craurus. The one is attended with pain and heaviness in the head, and this is the commoner of the two, the other with diarrhoea. The latter is incurable, the former is treated by applying wine fomentations to the snout and rinsing the nostrils with wine. Even this disease is very hard to cure; it has been known to kill within three or four days. The animal is chiefly subject to branchos when it gets extremely fat, and when the heat has brought a good supply of figs. The treatment is to feed on mashed mulberries, to give repeated warm baths, and to lance the under part of the tongue.

Pigs with flabby flesh are subject to measles about the legs, neck, and shoulders, for the pimples develop chiefly in these parts. If the pimples are few in number the flesh is comparatively sweet, but if they be numerous it gets watery and flaccid. The symptoms of measles are obvious, for the pimples show chiefly on the underside of the tongue, and if you pluck the bristles off the chine the skin will appear suffused with blood, and further the animal will be unable to keep its hind-feet at rest. Pigs never take this disease while they are mere sucklings. The pimples may be got rid of by feeding on this kind of spelt called tiphe; and this spelt, by the way, is very good for ordinary food. The best food for rearing and fattening pigs is chickpeas and figs, but the one thing essential is to vary the food as much as possible, for this animal, like animals in general lights in a change of diet; and it is said that one kind of food blows the animal out, that another superinduces flesh, and that another puts on fat, and that acorns, though liked by the animal, render the flesh flaccid. Besides, if a sow eats acorns in great quantities, it will miscarry, as is also the case with the ewe; and, indeed, the miscarriage is more certain in the case of the ewe than in the case of the sow. The pig is the only animal known to be subject to measles.

Part 22

Dogs suffer from three diseases; rabies, quinsy, and sore feet. Rabies drives the animal mad, and any animal whatever, excepting man, will take the disease if bitten by a dog so afflicted; the disease is fatal to the dog itself, and to any animal it may bite, man excepted. Quinsy also is fatal to dogs; and only a few recover from disease of the feet. The camel, like the dog, is subject to rabies. The elephant, which is reputed to enjoy immunity from all other illnesses, is occasionally subject to flatulency.

Part 23

Cattle in herds are liable to two diseases, foot, sickness and craurus. In the former their feet suffer from eruptions, but the animal recovers from the disease without even the loss of the hoof. It is found of service to smear the horny parts with warm pitch. In craurus, the breath comes warm at short intervals; in fact, craurus in cattle answers to fever in man. The symptoms of the disease are drooping of the ears and disinclination for food. The animal soon succumbs, and when the carcase is opened the lungs are found to be rotten.

Part 24

Horses out at pasture are free from all diseases excepting disease of the feet. From this disease they sometimes lose their hooves: but after losing them they grow them soon again, for as one hoof is decaying it is being replaced by another. Symptoms of the malady are a sinking in and wrinkling of the lip in the middle under the nostrils, and in the case of the male, a twitching of the right testicle.

Stall-reared horses are subject to very numerous forms of disease. They are liable to disease called 'eileus'. Under this disease the animal trails its hind-legs under its belly so far forward as almost to fall back on its haunches; if it goes without food for several days and turns rabid, it may be of service to draw blood, or to castrate the male. The animal is subject also to tetanus: the veins get rigid, as also the head and neck, and the animal walks with its legs stretched out straight. The horse suffers also from abscesses. Another painful illness afflicts them called the 'barley-surfeit'. There are a softening of the palate and heat of the breath; the animal may recover through the strength of its own constitution, but no formal remedies are of any avail.

There is also a disease called nymphia, in which the animal is said to stand still and droop its head on hearing flute-music; if during this ailment the horse be mounted, it will run off at a gallop until it is pulled. Even with this rabies in full force, it preserves a dejected spiritless appearance; some of the symptoms are a throwing back of the ears followed by a projection of them, great languor, and heavy breathing. Heart-ache also is incurable, of which the symptom is a drawing in of the flanks; and so is displacement of the bladder, which is accompanied by a retention of urine and a drawing up of the hooves and haunches. Neither is there any cure if the animal swallow the grape-beetle, which is about the size of the sphondyle or knuckle-beetle. The bite of the shrewmouse is dangerous to horses and other draught animals as well; it is followed by boils. The bite is all the more dangerous if the mouse be pregnant when she bites, for the boils then burst, but do not burst otherwise. The cicigna-called 'chalcis' by some, and 'zignis' by others-either causes death by its bite or, at all events, intense pain; it is like a small lizard, with the colour of the blind snake. In point of fact, according to experts, the horse and the sheep have pretty well as many ailments as the human species. The drug known under the name of 'sandarace' or realgar, is extremely injurious to a horse, and to all draught animals; it is given to the animal as a medicine in a solution of water, the liquid being filtered through a colander. The mare when pregnant apt to miscarry when disturbed by the odour of an extinguished candle; and a similar accident happens occasionally to women in their pregnancy. So much for the diseases of the horse.

The so-called hippomanes grows, as has stated, on the foal, and the mare nibbles it off as she licks and cleans the foal. All the curious stories connected with the hippomanes are due to old wives and to the venders of charms. What is called the 'polium' or foal's membrane, is, as all the accounts state, delivered by the mother before the foal appears.

A horse will recognize the neighing of any other horse with which it may have fought at any previous period. The horse delights in meadows and marshes, and likes to drink muddy water; in fact, if water be clear, the horse will trample in it to make it turbid, will then drink it, and afterwards will wallow in it. The animal is fond of water in every way, whether for drinking or for bathing purposes; and this explains the peculiar constitution of the hippopotamus or river-horse. In regard to water the ox is the opposite of the horse; for if the water be impure or cold, or mixed up with alien matter, it will refuse to drink it.

Part 25

The ass suffers chiefly from one particular disease which they call 'melis'. It arises first in the head, and a clammy humour runs down the nostrils, thick and red; if it stays in the head the animal may recover, but if it descends into the lungs the animal will die. Of all animals on its of its kind it is the least capable of enduring extreme cold, which circumstance will account for the fact that the animal is not found on the shores of the Euxine, nor in Scythia.

Part 26

Elephants suffer from flatulence, and when thus afflicted can void neither solid nor liquid residuum. If the elephant swallow earth-mould it suffers from relaxation; but if it go on taking it steadily, it will experience no harm. From time to time it takes to swallowing stones. It suffers also from diarrhoea: in this case they administer draughts of lukewarm water or dip its fodder in honey, and either one or the other prescription will prove a costive. When they suffer from insomnia, they will be restored to health if their shoulders be rubbed with salt, olive-oil, and warm water; when they have aches in their shoulders they will derive great benefit from the application of roast pork. Some elephants like olive oil, and others do not. If there is a bit of iron in the inside of an elephant it is said that it will pass out if the animal takes a drink of olive-oil; if the animal refuses olive-oil, they soak a root in the oil and give it the root to swallow. So much, then, for quadrupeds.

Part 27

Insects, as a general rule, thrive best in the time of year in which they come into being, especially if the season be moist and warm, as in spring.

In bee-hives are found creatures that do great damage to the combs; for instance, the grub that spins a web and ruins the honeycomb: it is called the 'cleros'. It engenders an insect like itself, of a spider-shape, and brings disease into the swarm. There is another insect resembling the moth, called by some the 'pyraustes', that flies about a lighted candle: this creature engenders a brood full of a fine down. It is never stung by a bee, and can only be got out of a hive by fumigation. A caterpillar also is engendered in hives, of a species nicknamed the teredo, or 'borer', with which creature the bee never interferes. Bees suffer most when flowers are covered with mildew, or in seasons of drought.

All insects, without exception, die if they be smeared over with oil; and they die all the more rapidly if you smear their head with the oil and lay them out in the sun.

Part 28

Variety in animal life may be produced by variety of locality: thus in one place an animal will not be found at all, in another it will be small, or short-lived, or will not thrive. Sometimes this sort of difference is observed in closely adjacent districts. Thus, in the territory of Miletus, in one district cicadas are found while there are none in the district close adjoining; and in Cephalenia there is a river on one side of which the cicada is found and not on the other. In Pordoselene there is a public road one side of which the weasel is found but not on the other. In Boeotia the mole is found in great abundance in the neighbourhood of Orchomenus, but there are none in Lebadia though it is in the immediate vicinity, and if a mole be transported from the one district to the other it will refuse to burrow in the soil. The hare cannot live in Ithaca if introduced there; in fact it will be found dead, turned towards the point of the beach where it was landed. The horseman-ant is not found in Sicily; the croaking frog has only recently appeared in the neighbourhood of Cyrene. In the whole of Libya there is neither wild boar, nor stag, nor wild goat; and in India, according to Ctesias-no very good authority, by the way-there are no swine, wild or tame, but animals that are devoid of blood and such as go into hiding or go torpid are all of immense size there. In the Euxine there are no small molluscs nor testaceans, except a few here and there; but in the Red Sea all the testaceans are exceedingly large. In Syria the sheep have tails a cubit in breadth; the goats have ears a span and a palm long, and some have ears that flap down to the ground; and the cattle have humps on their shoulders, like the camel. In Lycia goats are shorn for their fleece, just as sheep are in all other countries. In Libya the long-horned ram is born with horns, and not the ram only, as Homer' words it, but the ewe as well; in Pontus, on the confines of Scythia, the ram is without horns.

In Egypt animals, as a rule, are larger than their congeners in Greece, as the cow and the sheep; but some are less, as the dog, the wolf, the hare, the fox, the raven, and the hawk; others are of pretty much the same size, as the crow and the goat. The difference, where it exists, is attributed to the food, as being abundant in one case and insufficient in another, for instance for the wolf and the hawk; for provision is scanty for the carnivorous animals, small birds being scarce; food is scanty also for the hare and for all frugivorous animals, because neither the nuts nor the fruit last long.

In many places the climate will account for peculiarities; thus in Illyria, Thrace, and Epirus the ass is small, and in Gaul and in Scythia the ass is not found at all owing to the coldness of the climate of these countries. In Arabia the lizard is more than a cubit in length, and the mouse is much larger than our field-mouse, with its hind-legs a span long and its front legs the length of the first finger-joint. In Libya, according to all accounts, the length of the serpents is something appalling; sailors spin a yarn to the effect that some crews once put ashore and saw the bones of a number of oxen, and that they were sure that the oxen had been devoured by serpents, for, just as they were putting out to sea, serpents came chasing their galleys at full speed and overturned one galley and set upon the crew. Again, lions are more numerous in Libya, and in that district of Europe that lies between the Achelous and the Nessus; the leopard is more abundant in Asia Minor, and is not found in Europe at all. As a general rule, wild animals are at their wildest in Asia, at their boldest in Europe, and most diverse in form in Libya; in fact, there is an old saying, 'Always something fresh in Libya.'

It would appear that in that country animals of diverse species meet, on account of the rainless climate, at the watering-places, and there pair together; and that such pairs will often breed if they be nearly of the same size and have periods of gestation of the same length. For it is said that they are tamed down in their behaviour towards each other by extremity of thirst. And, by the way, unlike animals elsewhere, they require to drink more in wintertime than in summer: for they acquire the habit of not drinking in summer, owing to the circumstance that there is usually no water then; and the mice, if they drink, die. Elsewhere also bastard-animals are born to heterogeneous pairs; thus in Cyrene the wolf and the bitch will couple and breed; and the Laconian hound is a cross between the fox and the dog. They say that the Indian dog is a cross between the tiger and the bitch, not the first cross, but a cross in the third generation; for they say that the first cross is a savage creature. They take the bitch to a lonely spot and tie her up: if the tiger be in an amorous mood he will pair with her; if not he will eat her up, and this casualty is of frequent occurrence.

Part 29

Locality will differentiate habits also: for instance, rugged highlands will not produce the same results as the soft lowlands. The animals of the highlands look fiercer and bolder, as is seen in the swine of Mount Athos; for a lowland boar is no match even for a mountain sow.

Again, locality is an important element in regard to the bite of an animal. Thus, in Pharos and other places, the bite of the scorpion is not dangerous; elsewhere-in Caria, for instances-where scorpions are venomous as well as plentiful and of large size, the sting is fatal to man or beast, even to the pig, and especially to a black pig, though the pig, by the way, is in general most singularly indifferent to the bite of any other creature. If a pig goes into water after being struck by the scorpion of Caria, it will surely die.

There is great variety in the effects produced by the bites of serpents. The asp is found in Libya; the so-called 'septic' drug is made from the body of the animal, and is the only remedy known for the bite of the original. Among the silphium, also, a snake is found, for the bite or which a certain stone is said to be a cure: a stone that is brought from the grave of an ancient king, which stone is put into water and drunk off. In certain parts of Italy the bite of the gecko is fatal. But the deadliest of all bites of venomous creatures is when one venomous animal has bitten another; as, for instance, a viper's after it has bitten a scorpion. To the great majority of such creatures man's is fatal. There is a very little snake, by some entitled the 'holy-snake', which is dreaded by even the largest serpents. It is about an ell long, and hairy-looking; whenever it bites an animal, the flesh all round the wound will at once mortify. There is in India a small snake which is exceptional in this respect, that for its bite no specific whatever is known.

Part 30

Animals also vary as to their condition of health in connexion with their pregnancy.

Testaceans, such as scallops and all the oyster-family, and crustaceans, such as the lobster family, are best when with spawn. Even in the case of the testacean we speak of spawning (or pregnancy); but whereas the crustaceans may be seen coupling and laying their spawn, this is never the case with testaceans. Molluscs are best in the breeding time, as the calamary, the sepia, and the octopus.

Fish, when they begin to breed, are nearly all good for the table; but after the female has gone long with spawn they are good in some cases, and in others are out of season. The maenis, for instance, is good at the breeding time. The female of this fish is round, the male longer and flatter; when the female is beginning to breed the male turns black and mottled, and is quite unfit for the table; at this period he is nicknamed the 'goat'.

The wrasses called the owzel and the thrush, and the smaris have different colours at different seasons, as is the case with the plumage of certain birds; that is to say, they become black in the spring and after the spring get white again. The phycis also changes its hue: in general it is white, but in spring it is mottled; it is the only sea-fish which is said make a bed for itself, and the female lays her spawn in this bed or nest. The maenis, as was observed, changes its colour as does the smaris, and in summer-time changes back from whitish to black, the change being especially marked about the fins and gills. The coracine, like the maenis, is in best condition at breeding time; the mullet, the bass, and scaly fishes in general are in bad condition at this period. A few fish are in much the same condition at all times, whether with spawn or not, as the glaucus. Old fish also are bad eating; the old tunny is unfit even for pickling, as a great part of its flesh wastes away with age, and the same wasting is observed in all old fishes. The age of a scaly fish may be told by the size and the hardness of its scales. An old tunny has been caught weighing fifteen talents, with the span of its tail two cubits and a palm broad.

River-fish and lake-fish are best after they have discharged the spawn in the case of the female and the milt in the case of the male: that is, when they have fully recovered from the exhaustion of such discharge. Some are good in the breeding time, as the saperdis, and some bad, as the sheat-fish. As a general rule, the male fish is better eating than the female; but the reverse holds good of the sheat-fish. The eels that are called females are the best for the table: they look as though they were female, but they really are not so.

Book IX

Part 1

Of the animals that are comparatively obscure and short-lived the characters or dispositions are not so obvious to recognition as are those of animals that are longer-lived. These latter animals appear to have a natural capacity corresponding to each of the passions: to cunning or simplicity, courage or timidity, to good temper or to bad, and to other similar dispositions of mind.

Some also are capable of giving or receiving instruction-of receiving it from one another or from man: those that have the faculty of hearing, for instance; and, not to limit the matter to audible sound, such as can differentiate the suggested meanings of word and gesture.

In all genera in which the distinction of male and female is found, Nature makes a similar differentiation in the mental characteristics of the two sexes. This differentiation is the most obvious in the case of human kind and in that of the larger animals and the viviparous quadrupeds. In the case of these latter the female softer in character, is the sooner tamed, admits more readily of caressing, is more apt in the way of learning; as, for instance, in the Laconian breed of dogs the female is cleverer than the male. Of the Molossian breed of dogs, such as are employed in the chase are pretty much the same as those elsewhere; but sheep-dogs of this breed are superior to the others in size, and in the courage with which they face the attacks of wild animals.

Dogs that are born of a mixed breed between these two kinds are remarkable for courage and endurance of hard labour.

In all cases, excepting those of the bear and leopard, the female is less spirited than the male; in regard to the two exceptional cases, the superiority in courage rests with the female. With all other animals the female is softer in disposition than the male, is more mischievous, less simple, more impulsive, and more attentive to the nurture of the young: the male, on the other hand, is more spirited than the female, more savage, more simple and less cunning. The traces of these differentiated characteristics are more or less visible everywhere, but they are especially visible where character is the more developed, and most of all in man.

The fact is, the nature of man is the most rounded off and complete, and consequently in man the qualities or capacities above referred to are found in their perfection. Hence woman is more compassionate than man, more easily moved to tears, at the same time is more jealous, more querulous, more apt to scold and to strike. She is, furthermore, more prone to despondency and less hopeful than the man, more void of shame or self-respect, more false of speech, more deceptive, and of more retentive memory. She is also more wakeful, more shrinking, more difficult to rouse to action, and requires a smaller quantity of nutriment.

As was previously stated, the male is more courageous than the female, and more sympathetic in the way of standing by to help. Even in the case of molluscs, when the cuttle-fish is struck with the trident the male stands by to help the female; but when the male is struck the female runs away.

There is enmity between such animals as dwell in the same localities or subsist on the food. If the means of subsistence run short, creatures of like kind will fight together. Thus it is said that seals which inhabit one and the same district will fight, male with male, and female with female, until one combatant kills the other, or one is driven away by the other; and their young do even in like manner.

All creatures are at enmity with the carnivores, and the carnivores with all the rest, for they all subsist on living creatures. Soothsayers take notice of cases where animals keep apart from one another, and cases where they congregate together; calling those that live at war with one another 'dissociates', and those that dwell in peace with one another 'associates'. One may go so far as to say that if there were no lack or stint of food, then those animals that are now afraid of man or are wild by nature would be tame and familiar with him, and in like manner with one another. This is shown by the way animals are treated in Egypt, for owing to the fact that food is constantly supplied to them the very fiercest creatures live peaceably together. The fact is they are tamed by kindness, and in some places crocodiles are tame to their priestly keeper from being fed by him. And elsewhere also the same phenomenon is to be observed.

The eagle and the snake are enemies, for the eagle lives on snakes; so are the ichneumon and the venom-spider, for the ichneumon preys upon the latter. In the case of birds, there is mutual enmity between the poecilis, the crested lark, the woodpecker (?), and the chloreus, for they devour one another's eggs; so also between the crow and the owl; for, owing to the fact that the owl is dim-sighted by day, the crow at midday preys upon the owl's eggs, and the owl at night upon the crow's, each having the whip-hand of the other, turn and turn about, night and day.

There is enmity also between the owl and the wren; for the latter also devours the owl's eggs. In the daytime all other little birds flutter round the owl-a practice which is popularly termed 'admiring him'-buffet him, and pluck out his feathers; in consequence of this habit, bird-catchers use the owl as a decoy for catching little birds of all kinds.

The so-called presbys or 'old man' is at war with the weasel and the crow, for they prey on her eggs and her brood; and so the turtle-dove with the pyrallis, for they live in the same districts and on the same food; and so with the green wood pecker and the libyus; and so with kite and the raven, for, owing to his having the advantage from stronger talons and more rapid flight the former can steal whatever the latter is holding, so that it is food also that makes enemies of these. In like manner there is war between birds that get their living from the sea, as between the brenthus, the gull, and the harpe; and so between the buzzard on one side and the toad and snake on the other, for the buzzard preys upon the eggs of the two others; and so between the turtle-dove and the chloreus; the chloreus kills the dove, and the crow kills the so-called drummer-bird.

The aegolius, and birds of prey in general, prey upon the calaris, and consequently there is war between it and them; and so is there war between the gecko-lizard and the spider, for the former preys upon the latter; and so between the woodpecker and the heron, for the former preys upon the eggs and brood of the latter. And so between the aegithus and the ass, owing to the fact that the ass, in passing a furze-bush, rubs its sore and itching parts against the prickles; by so doing, and all the more if it brays, it topples the eggs and the brood out of the nest, the young ones tumble out in fright, and the mother-bird, to avenge this wrong, flies at the beast and pecks at his sore places.

The wolf is at war with the ass, the bull, and the fox, for as being a carnivore, he attacks these other animals; and so for the same reason with the fox and the circus, for the circus, being carnivorous and furnished with crooked talons, attacks and maims the animal. And so the raven is at war with the bull and the ass, for it flies at them, and strikes them, and pecks at their eyes; and so with the eagle and the heron, for the former, having crooked talons, attacks the latter, and the latter usually succumbs to the attack; and so the merlin with the vulture; and the crex with the eleus-owl, the blackbird, and the oriole (of this latter bird, by the way, the story goes that he was originally born out of a funeral pyre): the cause of warfare is that the crex injures both them and their young. The nuthatch and the wren are at war with the eagle; the nuthatch breaks the eagle's eggs, so the eagle is at war with it on special grounds, though, as a bird of prey, it carries on a general war all round. The horse and the anthus are enemies, and the horse will drive the bird out of the field where he is grazing: the bird feeds on grass, and sees too dimly to foresee an attack; it mimics the whinnying of the horse, flies at him, and tries to frighten him away; but the horse drives the bird away, and whenever he catches it he kills it: this bird lives beside rivers or on marsh ground; it has pretty plumage, and finds its without trouble. The ass is at enmity with the lizard, for the lizard sleeps in his manger, gets into his nostril, and prevents his eating.

Of herons there are three kinds: the ash coloured, the white, and the starry heron (or bittern). Of these the first mentioned submits with reluctance to the duties of incubation, or to union of the sexes; in fact, it screams during the union, and it is said drips blood from its eyes; it lays its eggs also in an awkward manner, not unattended with pain. It is at war with certain creatures that do it injury: with the eagle for robbing it, with the fox for worrying it at night, and with the lark for stealing its eggs.

The snake is at war with the weasel and the pig; with the weasel when they are both at home, for they live on the same food; with the pig for preying on her kind. The merlin is at war with the fox; it strikes and claws it, and, as it has crooked talons, it kills the animal's young. The raven and the fox are good friends, for the raven is at enmity with the merlin; and so when the merlin assails the fox the raven comes and helps the animal. The vulture and the merlin are mutual enemies, as being both furnished with crooked talons. The vulture fights with the eagle, and so, by the way, does the swan; and the swan is often victorious: moreover, of all birds swans are most prone to the killing of one another.

In regard to wild creatures, some sets are at enmity with other sets at all times and under all circumstances; others, as in the case of man and man, at special times and under incidental circumstances. The ass and the acanthis are enemies; for the bird lives on thistles, and the ass browses on thistles when they are young and tender. The anthus, the acanthis, and the aegithus are at enmity with one another; it is said that the blood of the anthus will not intercommingle with the blood of the aegithus. The crow and the heron are friends, as also are the sedge-bird and lark, the laedus and the celeus or green woodpecker; the woodpecker lives on the banks of rivers and beside brakes, the laedus lives on rocks and bills, and is greatly attached to its nesting-place. The piphinx, the harpe, and the kite are friends; as are the fox and the snake, for both burrow underground; so also are the blackbird and the turtle-dove. The lion and the thos or civet are enemies, for both are carnivorous and live on the same food. Elephants fight fiercely with one another, and stab one another with their tusks; of two combatants the beaten one gets completely cowed, and dreads the sound of his conqueror's voice. These animals differ from one another an extraordinary extent in the way of courage. Indians employ these animals for war purposes, irrespective of sex; the females, however, are less in size and much inferior in point of spirit. An elephant by pushing with his big tusks can batter down a wall, and will butt with his forehead at a palm until he brings it down, when he stamps on it and lays it in orderly fashion on the ground. Men hunt the elephant in the following way: they mount tame elephants of approved spirit and proceed in quest of wild animals; when they come up with these they bid the tame brutes to beat the wild ones until they tire the latter completely. Hereupon the driver mounts a wild brute and guides him with the application of his metal prong; after this the creature soon becomes tame, and obeys guidance. Now when the driver is on their back they are all tractable, but after he has dismounted, some are tame and others vicious; in the case of these latter, they tie their front-legs with ropes to keep them quiet. The animal is hunted whether young or full grown.

Thus we see that in the case of the creatures above mentioned their mutual friendship or the is due to the food they feed on and the life they lead.

Part 2

Of fish, such as swim in shoals together are friendly to one another; such as do not so swim are enemies. Some fishes swarm during the spawning season; others after they have spawned. To state the matter comprehensively, we may say that the following are shoaling fish: the tunny, the maenis, the sea-gudgeon, the bogue, the horse-mackerel, the coracine, the synodon or dentex, the red mullet, the sphyraena, the anthias, the eleginus, the atherine, the sarginus, the gar-fish, (the squid,) the rainbow-wrasse, the pelamyd, the mackerel, the coly-mackerel. Of these some not only swim in shoals, but go in pairs inside the shoal; the rest without exception swim in pairs, and only swim in shoals at certain periods: that is, as has been said, when they are heavy with spawn or after they have spawned.

The bass and the grey mullet are bitter enemies, but they swarm together at certain times; for at times not only do fishes of the same species swarm together, but also those whose feeding-grounds are identical or adjacent, if the food-supply be abundant. The grey mullet is often found alive with its tail lopped off, and the conger with all that part of its body removed that lies to the rear of the vent; in the case of the mullet the injury is wrought by the bass, in that of the conger-eel by the muraena. There is war between the larger and the lesser fishes: for the big fishes prey on the little ones. So much on the subject of marine animals.

Part 3

The characters of animals, as has been observed, differ in respect to timidity, to gentleness, to courage, to tameness, to intelligence, and to stupidity.

The sheep is said to be naturally dull and stupid. Of all quadrupeds it is the most foolish: it will saunter away to lonely places with no object in view; oftentimes in stormy weather it will stray from shelter; if it be overtaken by a snowstorm, it will stand still unless the shepherd sets it in motion; it will stay behind and perish unless the shepherd brings up the rams; it will then follow home.

If you catch hold of a goat's beard at the extremity-the beard is of a substance resembling hair-all the companion goats will stand stock still, staring at this particular goat in a kind of dumbfounderment.

You will have a warmer bed in amongst the goats than among the sheep, because the goats will be quieter and will creep up towards you; for the goat is more impatient of cold than the sheep.

Shepherds train sheep to close in together at a clap of their hands, for if, when a thunderstorm comes on, a ewe stays behind without closing in, the storm will kill it if it be with young; consequently if a sudden clap or noise is made, they close in together within the sheepfold by reason of their training.

Even bulls, when they are roaming by themselves apart from the herd, are killed by wild animals.

Sheep and goats lie crowded together, kin by kin. When the sun turns early towards its setting, the goats are said to lie no longer face to face, but back to back.

Part 4

Cattle at pasture keep together in their accustomed herds, and if one animal strays away the rest will follow; consequently if the herdsmen lose one particular animal, they keep close watch on all the rest.

When mares with their colts pasture together in the same field, if one dam dies the others will take up the rearing of the colt. In point of fact, the mare appears to be singularly prone by nature to maternal fondness; in proof whereof a barren mare will steal the foal from its dam, will tend it with all the solicitude of a mother, but, as it will be unprovided with mother's milk, its solicitude will prove fatal to its charge.

Part 5

Among wild quadrupeds the hind appears to be pre-eminently intelligent; for example, in its habit of bringing forth its young on the sides of public roads, where the fear of man forbids the approach of wild animals. Again, after parturition, it first swallows the afterbirth, then goes in quest of the seseli shrub, and after eating of it returns to its young. The mother takes its young betimes to her lair, so leading it to know its place of refuge in time of danger; this lair is a precipitous rock, with only one approach, and there it is said to hold its own against all comers. The male when it gets fat, which it does in a high degree in autumn, disappears, abandoning its usual resorts, apparently under an idea that its fatness facilitates its capture. They shed their horns in places difficult of access or discovery, whence the proverbial expression of 'the place where the stag sheds his horns'; the fact being that, as having parted with their weapons, they take care not to be seen. The saying is that no man has ever seen the animal's left horn; that the creature keeps it out of sight because it possesses some medicinal property.

In their first year stags grow no horns, but only an excrescence indicating where horns will be, this excrescence being short and thick. In their second year they grow their horns for the first time, straight in shape, like pegs for hanging clothes on; and on this account they have an appropriate nickname. In the third year the antlers are bifurcate; in the fourth year they grow trifurcate; and so they go on increasing in complexity until the creature is six years old: after this they grow their horns without any specific differentiation, so that you cannot by observation of them tell the animal's age. But the patriarchs of the herd may be told chiefly by two signs; in the first place they have few teeth or none at all, and, in the second place, they have ceased to grow the pointed tips to their antlers. The forward-pointing tips of the growing horns (that is to say the brow antlers), with which the animal meets attack, are technically termed its 'defenders'; with these the patriarchs are unprovided, and their antlers merely grow straight upwards. Stags shed their horns annually, in or about the month of May; after shedding, they conceal themselves, it is said, during the daytime, and, to avoid the flies, hide in thick copses; during this time, until they have grown their horns, they feed at night-time. The horns at first grow in a kind of skin envelope, and get rough by degrees; when they reach their full size the animal basks in the sun, to mature and dry them. When they need no longer rub them against tree-trunks they quit their hiding places, from a sense of security based upon the possession of arms defensive and offensive. An Achaeine stag has been caught with a quantity of green ivy grown over its horns, it having grown apparently, as on fresh green wood, when the horns were young and tender. When a stag is stung by a venom-spider or similar insect, it gathers crabs and eats them; it is said to be a good thing for man to drink the juice, but the taste is disagreeable. The hinds after parturition at once swallow the afterbirth, and it is impossible to secure it, for the hind catches it before it falls to the ground: now this substance is supposed to have medicinal properties. When hunted the creatures are caught by singing or pipe-playing on the part of the hunters; they are so pleased with the music that they lie down on the grass. If there be two hunters, one before their eyes sings or plays the pipe, the other keeps out of sight and shoots, at a signal given by the confederate. If the animal has its ears cocked, it can hear well and you cannot escape its ken; if its ears are down, you can.

Part 6

When bears are running away from their pursuers they push their cubs in front of them, or take them up and carry them; when they are being overtaken they climb up a tree. When emerging from their winter-den, they at once take to eating cuckoo-pint, as has been said, and chew sticks of wood as though they were cutting teeth.

Many other quadrupeds help themselves in clever ways. Wild goats in Crete are said, when wounded by arrows, to go in search of dittany, which is supposed to have the property of ejecting arrows in the body. Dogs, when they are ill, eat some kind of grass and produce vomiting. The panther, after eating panther's-bane, tries to find some human excrement, which is said to heal its pain. This panther's-bane kills lions as well. Hunters hang up human excrement in a vessel attached to the boughs of a tree, to keep the animal from straying to any distance; the animal meets its end in leaping up to the branch and trying to get at the medicine. They say that the panther has found out that wild animals are fond of the scent it emits; that, when it goes a-hunting, it hides itself; that the other animals come nearer and nearer, and that by this stratagem it can catch even animals as swift of foot as stags.

The Egyptian ichneumon, when it sees the serpent called the asp, does not attack it until it has called in other ichneumons to help; to meet the blows and bites of their enemy the assailants beplaster themselves with mud, by first soaking in the river and then rolling on the ground.

When the crocodile yawns, the trochilus flies into his mouth and cleans his teeth. The trochilus gets his food thereby, and the crocodile gets ease and comfort; it makes no attempt to injure its little friend, but, when it wants it to go, it shakes its neck in warning, lest it should accidentally bite the bird.

The tortoise, when it has partaken of a snake, eats marjoram; this action has been actually observed. A man saw a tortoise perform this operation over and over again, and every time it plucked up some marjoram go back to partake of its prey; he thereupon pulled the marjoram up by the roots, and the consequence was the tortoise died. The weasel, when it fights with a snake, first eats wild rue, the smell of which is noxious to the snake. The dragon, when it eats fruit, swallows endive-juice; it has been seen in the act. Dogs, when they suffer from worms, eat the standing corn. Storks, and all other birds, when they get a wound fighting, apply marjoram to the place injured.

Many have seen the locust, when fighting with the snake get a tight hold of the snake by the neck. The weasel has a clever way of getting the better of birds; it tears their throats open, as wolves do with sheep. Weasels fight desperately with mice-catching snakes, as they both prey on the same animal.

In regard to the instinct of hedgehogs, it has been observed in many places that, when the wind is shifting from north to south, and from south to north, they shift the outlook of their earth-holes, and those that are kept in domestication shift over from one wall to the other. The story goes that a man in Byzantium got into high repute for foretelling a change of weather, all owing to his having noticed this habit of the hedgehog.

The polecat or marten is about as large as the smaller breed of Maltese dogs. In the thickness of its fur, in its look, in the white of its belly, and in its love of mischief, it resembles the weasel; it is easily tamed; from its liking for honey it is a plague to bee-hives; it preys on birds like the cat. Its genital organ, as has been said, consists of bone: the organ of the male is supposed to be a cure for strangury; doctors scrape it into powder, and administer it in that form.

Part 7

In a general way in the lives of animals many resemblances to human life may be observed. Pre-eminent intelligence will be seen more in small creatures than in large ones, as is exemplified in the case of birds by the nest building of the swallow. In the same way as men do, the bird mixes mud and chaff together; if it runs short of mud, it souses its body in water and rolls about in the dry dust with wet feathers; furthermore, just as man does, it makes a bed of straw, putting hard material below for a foundation, and adapting all to suit its own size. Both parents co-operate in the rearing of the young; each of the parents will detect, with practised eye, the young one that has had a helping, and will take care it is not helped twice over; at first the parents will rid the nest of excrement, but, when the young are grown, they will teach their young to shift their position and let their excrement fall over the side of the nest.

Pigeons exhibit other phenomena with a similar likeness to the ways of humankind. In pairing the same male and the same female keep together; and the union is only broken by the death of one of the two parties. At the time of parturition in the female the sympathetic attentions of the male are extraordinary; if the female is afraid on account of the impending parturition to enter the nest, the male will beat her and force her to come in. When the young are born, he will take and masticate pieces of suitable food, will open the beaks of the fledglings, and inject these pieces, thus preparing them betimes to take food. (When the male bird is about to expel the the young ones from the nest he cohabits with them all.) As a general rule these birds show this conjugal fidelity, but occasionally a female will cohabit with other than her mate. These birds are combative, and quarrel with one another, and enter each other's nests, though this occurs but seldom; at a distance from their nests this quarrelsomeness is less marked, but in the close neighbourhood of their nests they will fight desperately. A peculiarity common to the tame pigeon, the ring-dove and the turtle-dove is that they do not lean the head back when they are in the act of drinking, but only when they have fully quenched their thirst. The turtle-dove and the ring-dove both have but one mate, and let no other come nigh; both sexes co-operate in the process of incubation. It is difficult to distinguish between the sexes except by an examination of their interiors. Ring-doves are long-lived; cases have been known where such birds were twenty-five years old, thirty years old, and in some cases forty. As they grow old their claws increase in size, and pigeon-fanciers cut the claws; as far as one can see, the birds suffer no other perceptible disfigurement by their increase in age. Turtle-doves and pigeons that are blinded by fanciers for use as decoys, live for eight years. Partridges live for about fifteen years. Ring-doves and turtle-doves always build their nests in the same place year after year. The male, as a general rule, is more long-lived than the female; but in the case of pigeons some assert that the male dies before the female, taking their inference from the statements of persons who keep decoy-birds in captivity. Some declare that the male sparrow lives only a year, pointing to the fact that early in spring the male sparrow has no black beard, but has one later on, as though the blackbearded birds of the last year had all died out; they also say that the females are the longer lived, on the grounds that they are caught in amongst the young birds and that their age is rendered manifest by the hardness about their beaks. Turtle-doves in summer live in cold places, (and in warm places during the winter); chaffinches affect warm habitations in summer and cold ones in winter.

Part 8

Birds of a heavy build, such as quails, partridges, and the like, build no nests; indeed, where they are incapable of flight, it would be of no use if they could do so. After scraping a hole on a level piece of ground-and it is only in such a place that they lay their eggs-they cover it over with thorns and sticks for security against hawks and eagles, and there lay their eggs and hatch them; after the hatching is over, they at once lead the young out from the nest, as they are not able to fly afield for food for them. Quails and partridges, like barn-door hens, when they go to rest, gather their brood under their wings. Not to be discovered, as might be the case if they stayed long in one spot, they do not hatch the eggs where they laid them. When a man comes by chance upon a young brood, and tries to catch them, the hen-bird rolls in front of the hunter, pretending to be lame: the man every moment thinks he is on the point of catching her, and so she draws him on and on, until every one of her brood has had time to escape; hereupon she returns to the nest and calls the young back. The partridge lays not less than ten eggs, and often lays as many as sixteen. As has been observed, the bird has mischievous and deceitful habits. In the spring-time, a noisy scrimmage takes place, out of which the male-birds emerge each with a hen. Owing to the lecherous nature of the bird, and from a dislike to the hen sitting, the males, if they find any eggs, roll them over and over until they break them in pieces; to provide against this the female goes to a distance and lays the eggs, and often, under the stress of parturition, lays them in any chance spot that offers; if the male be near at hand, then to keep the eggs intact she refrains from visiting them. If she be seen by a man, then, just as with her fledged brood, she entices him off by showing herself close at his feet until she has drawn him to a distance. When the females have run away and taken to sitting, the males in a pack take to screaming and fighting; when thus engaged, they have the nickname of 'widowers'. The bird who is beaten follows his victor, and submits to be covered by him only; and the beaten bird is covered by a second one or by any other, only clandestinely without the victor's knowledge; this is so, not at all times, but at a particular season of the year, and with quails as well as with partridges. A similar proceeding takes place occasionally with barn-door cocks: for in temples, where cocks are set apart as dedicate without hens, they all as a matter of course tread any new-comer. Tame partridges tread wild birds, pecket their heads, and treat them with every possible outrage. The leader of the wild birds, with a counter-note of challenge, pushes forward to attack the decoy-bird, and after he has been netted, another advances with a similar note. This is what is done if the decoy be a male; but if it be a female that is the decoy and gives the note, and the leader of the wild birds give a counter one, the rest of the males set upon him and chase him away from the female for making advances to her instead of to them; in consequence of this the male often advances without uttering any cry, so that no other may hear him and come and give him battle; and experienced fowlers assert that sometimes the male bird, when he approaches the female, makes her keep silence, to avoid having to give battle to other males who might have heard him. The partridge has not only the note here referred to, but also a thin shrill cry and other notes. Oftentimes the hen-bird rises from off her brood when she sees the male showing attentions to the female decoy; she will give the counter note and remain still, so as to be trodden by him and divert him from the decoy. The quail and the partridge are so intent upon sexual union that they often come right in the way of the decoy-birds, and not seldom alight upon their heads. So much for the sexual proclivities of the partridge, for the way in which it is hunted, and the general nasty habits of the bird.

As has been said, quails and partridges build their nests upon the ground, and so also do some of the birds that are capable of sustained flight. Further, for instance, of such birds, the lark and the woodcock, as well as the quail, do not perch on a branch, but squat upon the ground.

Part 9

The woodpecker does not squat on the ground, but pecks at the bark of trees to drive out from under it maggots and gnats; when they emerge, it licks them up with its tongue, which is large and flat. It can run up and down a tree in any way, even with the head downwards, like the gecko-lizard. For secure hold upon a tree, its claws are better adapted than those of the daw; it makes its way by sticking these claws into the bark. One species of woodpecker is smaller than a blackbird, and has small reddish speckles; a second species is larger than the blackbird, and a third is not much smaller than a barn-door hen. It builds a nest on trees, as has been said, on olive trees amongst others. It feeds on the maggots and ants that are under the bark: it is so eager in the search for maggots that it is said sometimes to hollow a tree out to its downfall. A woodpecker once, in course of domestication, was seen to insert an almond into a hole in a piece of timber, so that it might remain steady under its pecking; at the third peck it split the shell of the fruit, and then ate the kernel.

Part 10

Many indications of high intelligence are given by cranes. They will fly to a great distance and up in the air, to command an extensive view; if they see clouds and signs of bad weather they fly down again and remain still. They, furthermore, have a leader in their flight, and patrols that scream on the confines of the flock so as to be heard by all. When they settle down, the main body go to sleep with their heads under their wing, standing first on one leg and then on the other, while their leader, with his head uncovered, keeps a sharp look out, and when he sees anything of importance signals it with a cry.

Pelicans that live beside rivers swallow the large smooth mussel-shells: after cooking them inside the crop that precedes the stomach, they spit them out, so that, now when their shells are open, they may pick the flesh out and eat it.

Part 11

Of wild birds, the nests are fashioned to meet the exigencies of existence and ensure the security of the young. Some of these birds are fond of their young and take great care of them, others are quite the reverse; some are clever in procuring subsistence, others are not so. Some of these birds build in ravines and clefts, and on cliffs, as, for instance, the so-called charadrius, or stone-curlew; this bird is in no way noteworthy for plumage or voice; it makes an appearance at night, but in the daytime keeps out of sight.

The hawk also builds in inaccessible places. Although a ravenous bird, it will never eat the heart of any bird it catches; this has been observed in the case of the quail, the thrush, and other birds. They modify betimes their method of hunting, for in summer they do not grab their prey as they do at other seasons.

Of the vulture, it is said that no one has ever seen either its young or its nest; on this account and on the ground that all of a sudden great numbers of them will appear without any one being able to tell from whence they come, Herodorus, the father of Bryson the sophist, says that it belongs to some distant and elevated land. The reason is that the bird has its nest on inaccessible crags, and is found only in a few localities. The female lays one egg as a rule, and two at the most.

Some birds live on mountains or in forests, as the hoopoe and the brenthus; this latter bird finds his food with ease and has a musical voice. The wren lives in brakes and crevices; it is difficult of capture, keeps out of sight, is gentle of disposition, finds its food with ease, and is something of a mechanic. It goes by the nickname of 'old man' or 'king'; and the story goes that for this reason the eagle is at war with him.

Part 12

Some birds live on the sea-shore, as the wagtail; the bird is of a mischievous nature, hard to capture, but when caught capable of complete domestication; it is a cripple, as being weak in its hinder quarters.

Web-footed birds without exception live near the sea or rivers or pools, as they naturally resort to places adapted to their structure. Several birds, however, with cloven toes live near pools or marshes, as, for instance, the anthus lives by the side of rivers; the plumage of this bird is pretty, and it finds its food with ease. The catarrhactes lives near the sea; when it makes a dive, it will keep under water for as long as it would take a man to walk a furlong; it is less than the common hawk. Swans are web-footed, and live near pools and marshes; they find their food with ease, are good-tempered, are fond of their young, and live to a green old age. If the eagle attacks them they will repel the attack and get the better of their assailant, but they are never the first to attack. They are musical, and sing chiefly at the approach of death; at this time they fly out to sea, and men, when sailing past the coast of Libya, have fallen in with many of them out at sea singing in mournful strains, and have actually seen some of them dying.

The cymindis is seldom seen, as it lives on mountains; it is black in colour, and about the size of the hawk called the 'dove-killer'; it is long and slender in form. The Ionians call the bird by this name; Homer in the Iliad mentions it in the line:

Chalcis its name with those of heavenly birth, But called Cymindis by the sons of earth.

The hybris, said by some to be the same as the eagle-owl, is never seen by daylight, as it is dim-sighted, but during the night it hunts like the eagle; it will fight the eagle with such desperation that the two combatants are often captured alive by shepherds; it lays two eggs, and, like others we have mentioned, it builds on rocks and in caverns. Cranes also fight so desperately among themselves as to be caught when fighting, for they will not leave off; the crane lays two eggs.

Part 13

The jay has a great variety of notes: indeed, might almost say it had a different note for every day in the year. It lays about nine eggs; builds its nest on trees, out of hair and tags of wool; when acorns are getting scarce, it lays up a store of them in hiding.

It is a common story of the stork that the old birds are fed by their grateful progeny. Some tell a similar story of the bee-eater, and declare that the parents are fed by their young not only when growing old, but at an early period, as soon as the young are capable of feeding them; and the parent-birds stay inside the nest. The under part of the bird's wing is pale yellow; the upper part is dark blue, like that of the halcyon; the tips of the wings are About autumn-time it lays six or seven eggs, in overhanging banks where the soil is soft; there it burrows into the ground to a depth of six feet.

The greenfinch, so called from the colour of its belly, is as large as a lark; it lays four or five eggs, builds its nest out of the plant called comfrey, pulling it up by the roots, and makes an under-mattress to lie on of hair and wool. The blackbird and the jay build their nests after the same fashion. The nest of the penduline tit shows great mechanical skill; it has the appearance of a ball of flax, and the hole for entry is very small.

People who live where the bird comes from say that there exists a cinnamon bird which brings the cinnamon from some unknown localities, and builds its nest out of it; it builds on high trees on the slender top branches. They say that the inhabitants attach leaden weights to the tips of their arrows and therewith bring down the nests, and from the intertexture collect the cinnamon sticks.

Part 14

The halcyon is not much larger than the sparrow. Its colour is dark blue, green, and light purple; the whole body and wings, and especially parts about the neck, show these colours in a mixed way, without any colour being sharply defined; the beak is light green, long and slender: such, then, is the look of the bird. Its nest is like sea-balls, i.e. the things that by the name of halosachne or sea foam, only the colour is not the same. The colour of the nest is light red, and the shape is that of the long-necked gourd. The nests are larger than the largest sponge, though they vary in size; they are roofed over, and great part of them is solid and great part hollow. If you use a sharp knife it is not easy to cut the nest through; but if you cut it, and at the same time bruise it with your hand, it will soon crumble to pieces, like the halosachne. The opening is small, just enough for a tiny entrance, so that even if the nest upset the sea does not enter in; the hollow channels are like those in sponges. It is not known for certain of what material the nest is constructed; it is possibly made of the backbones of the gar-fish; for, by the way, the bird lives on fish. Besides living on the shore, it ascends fresh-water streams. It lays generally about five eggs, and lays eggs all its life long, beginning to do so at the age of four months.

Part 15

The hoopoe usually constructs its nest out of human excrement. It changes its appearance in summer and in winter, as in fact do the great majority of wild birds. (The titmouse is said to lay a very large quantity of eggs: next to the ostrich the blackheaded tit is said by some to lay the largest number of eggs; seventeen eggs have been seen; it lays, however, more than twenty; it is said always to lay an odd number. Like others we have mentioned, it builds in trees; it feeds on caterpillars.) A peculiarity of this bird and of the nightingale is that the outer extremity of the tongue is not sharp-pointed.

The aegithus finds its food with ease, has many young, and walks with a limp. The golden oriole is apt at learning, is clever at making a living, but is awkward in flight and has an ugly plumage.

Part 16

The reed-warbler makes its living as easily as any other bird, sits in summer in a shady spot facing the wind, in winter in a sunny and sheltered place among reeds in a marsh; it is small in size, with a pleasant note. The so-called chatterer has a pleasant note, beautiful plumage, makes a living cleverly, and is graceful in form; it appears to be alien to our country; at all events it is seldom seen at a distance from its own immediate home.

Part 17

The crake is quarrelsome, clever at making a living, but in other ways an unlucky bird. The bird called sitta is quarrelsome, but clever and tidy, makes its living with ease, and for its knowingness is regarded as uncanny; it has a numerous brood, of which it is fond, and lives by pecking the bark of trees. The aegolius-owl flies by night, is seldom seen by day; like others we have mentioned, it lives on cliffs or in caverns; it feeds on two kinds of food; it has a strong hold on life and is full of resource. The tree-creeper is a little bird, of fearless disposition; it lives among trees, feeds on caterpillars, makes a living with ease, and has a loud clear note. The acanthis finds its food with difficulty; its plumage is poor, but its note is musical.

Part 18

Of the herons, the ashen-coloured one, as has been said, unites with the female not without pain; it is full of resource, carries its food with it, is eager in the quest of it, and works by day; its plumage is poor, and its excrement is always wet. Of the other two species-for there are three in all-the white heron has handsome plumage, unites without harm to itself with the female, builds a nest and lays its eggs neatly in trees; it frequents marshes and lakes and Plains and meadow land. The speckled heron, which is nicknamed 'the skulker', is said in folklore stories to be of servile origin, and, as its nickname implies, it is the laziest bird of the three species. Such are the habits of herons. The bird that is called the poynx has this peculiarity, that it is more prone than any other bird to peck at the eyes of an assailant or its prey; it is at war with the harpy, as the two birds live on the same food.

Part 19

There are two kinds of owsels; the one is black, and is found everywhere, the other is quite white, about the same size as the other, and with the same pipe. This latter is found on Cyllene in Arcadia, and is found nowhere else. The laius, or blue-thrush, is like the black owsel, only a little smaller; it lives on cliffs or on tile roofings; it has not a red beak as the black owsel has.

Part 20

Of thrushes there are three species. One is the misselthrush; it feeds only on mistletoe and resin; it is about the size of the jay. A second is the song-thrush; it has a sharp pipe, and is about the size of the owsel. There is another species called the Illas; it is the smallest species of the three, and is less variegated in plumage than the others.

Part 21

There is a bird that lives on rocks, called the blue-bird from its colour. It is comparatively common in Nisyros, and is somewhat less than the owsel and a little bigger than the chaffinch. It has large claws, and climbs on the face of the rocks. It is steel-blue all over; its beak is long and slender; its legs are short, like those of the woodpecker.

Part 22

The oriole is yellow all over; it is not visible during winter, but puts in an appearance about the time of the summer solstice, and departs again at the rising of Arcturus; it is the size of the turtle-dove. The so-called soft-head (or shrike) always settles on one and the same branch, where it falls a prey to the birdcatcher. Its head is big, and composed of gristle; it is a little smaller than the thrush; its beak is strong, small, and round; it is ashen-coloured all over; is fleet of foot, but slow of wing. The bird-catcher usually catches it by help of the owl.

Part 23

There is also the pardalus. As a rule, it is seen in flocks and not singly; it is ashen-coloured all over, and about the size of the birds last described; it is fleet of foot and strong of wing, and its pipe is loud and high-pitched. The collyrion (or fieldfare) feeds on the same food as the owsel; is of the same size as the above mentioned birds; and is trapped usually in the winter. All these birds are found at all times. Further, there are the birds that live as a rule in towns, the raven and the crow. These also are visible at all seasons, never shift their place of abode, and never go into winter quarters.

Part 24

Of daws there are three species. One is the chough; it is as large as the crow, but has a red beak. There is another, called the 'wolf'; and further there is the little daw, called the 'railer'. There is another kind of daw found in Lybia and Phrygia, which is web-footed.

Part 25

Of larks there are two kinds. One lives on the ground and has a crest on its head; the other is gregarious, and not sporadic like the first; it is, however, of the same coloured plumage, but is smaller, and has no crest; it is an article of human food.

Part 26

The woodcock is caught with nets in gardens. It is about the size of a barn-door hen; it has a long beak, and in plumage is like the francolin-partridge. It runs quickly, and is pretty easily domesticated. The starling is speckled; it is of the same size as the owsel.

Part 27

Of the Egyptian ibis there are two kinds, the white and the black. The white ones are found over Egypt, excepting in Pelusium; the black ones are found in Pelusium, and nowhere else in Egypt.

Part 28

Of the little horned owls there are two kinds, and one is visible at all seasons, and for that reason has the nickname of 'all-the-year-round owl'; it is not sufficiently palatable to come to table; another species makes its appearance sometimes in the autumn, is seen for a single day or at the most for two days, and is regarded as a table delicacy; it scarcely differs from the first species save only in being fatter; it has no note, but the other species has. With regard to their origin, nothing is known from ocular observation; the only fact known for certain is that they are first seen when a west wind is blowing.

Part 29

The cuckoo, as has been said elsewhere, makes no nest, but deposits its eggs in an alien nest, generally in the nest of the ring-dove, or on the ground in the nest of the hypolais or lark, or on a tree in the nest of the green linnet. it lays only one egg and does not hatch it itself, but the mother-bird in whose nest it has deposited it hatches and rears it; and, as they say, this mother bird, when the young cuckoo has grown big, thrusts her own brood out of the nest and lets them perish; others say that this mother-bird kills her own brood and gives them to the alien to devour, despising her own young owing to the beauty of the cuckoo. Personal observers agree in telling most of these stories, but are not in agreement as to the instruction of the young. Some say that the mother-cuckoo comes and devours the brood of the rearing mother; others say that the young cuckoo from its superior size snaps up the food brought before the smaller brood have a chance, and that in consequence the smaller brood die of hunger; others say that, by its superior strength, it actually kills the other ones whilst it is being reared up with them. The cuckoo shows great sagacity in the disposal of its progeny; the fact is, the mother cuckoo is quite conscious of her own cowardice and of the fact that she could never help her young one in an emergency, and so, for the security of the young one, she makes of him a supposititious child in an alien nest. The truth is, this bird is pre-eminent among birds in the way of cowardice; it allows itself to be pecked at by little birds, and flies away from their attacks.

Part 30

It has already been stated that the footless bird, which some term the cypselus, resembles the swallow; indeed, it is not easy to distinguish between the two birds, excepting in the fact that the cypselus has feathers on the shank. These birds rear their young in long cells made of mud, and furnished with a hole just big enough for entry and exit; they build under cover of some roofing-under a rock or in a cavern-for protection against animals and men.

The so-called goat-sucker lives on mountains; it is a little larger than the owsel, and less than the cuckoo; it lays two eggs, or three at the most, and is of a sluggish disposition. It flies up to the she-goat and sucks its milk, from which habit it derives its name; it is said that, after it has sucked the teat of the animal, the teat dries up and the animal goes blind. It is dim-sighted in the day-time, but sees well enough by night.

Part 31

In narrow circumscribed districts where the food would be insufficient for more birds than two, ravens are only found in isolated pairs; when their young are old enough to fly, the parent couple first eject them from the nest, and by and by chase them from the neighbourhood. The raven lays four or five eggs. About the time when the mercenaries under Medius were slaughtered at Pharsalus, the districts about Athens and the Peloponnese were left destitute of ravens, from which it would appear that these birds have some means of intercommunicating with one another.

Part 32

Of eagles there are several species. One of them, called 'the white-tailed eagle', is found on low lands, in groves, and in the neighbourhood of cities; some call it the 'heron-killer'. It is bold enough to fly to mountains and the interior of forests. The other eagles seldom visit groves or low-lying land. There is another species called the 'plangus'; it ranks second in point of size and strength; it lives in mountain combes and glens, and by marshy lakes, and goes by the name of 'duck-killer' and 'swart-eagle.' It is mentioned by Homer in his account of the visit made by Priam to the tent of Achilles. There is another species with black Plumage, the smallest but boldest of all the kinds. It dwells on mountains or in forests, and is called 'the black-eagle' or 'the hare-killer'; it is the only eagle that rears its young and thoroughly takes them out with it. It is swift of flight, is neat and tidy in its habits, too proud for jealousy, fearless, quarrelsome; it is also silent, for it neither whimpers nor screams. There is another species, the percnopterus, very large, with white head, very short wings, long tail-feathers, in appearance like a vulture. It goes by the name of 'mountain-stork' or 'half-eagle'. It lives in groves; has all the bad qualities of the other species, and none of the good ones; for it lets itself be chased and caught by the raven and the other birds. It is clumsy in its movements, has difficulty in procuring its food, preys on dead animals, is always hungry, and at all times whining and screaming. There is another species, called the 'sea-eagle' or 'osprey'. This bird has a large thick neck, curved wings, and broad tailfeathers; it lives near the sea, grasps its prey with its talons, and often, from inability to carry it, tumbles down into the water. There is another species called the 'true-bred'; people say that these are the only true-bred birds to be found, that all other birds-eagles, hawks, and the smallest birds-are all spoilt by the interbreeding of different species. The true-bred eagle is the largest of all eagles; it is larger than the phene; is half as large again as the ordinary eagle, and has yellow plumage; it is seldom seen, as is the case with the so-called cymindis. The time for an eagle to be on the wing in search of prey is from midday to evening; in the morning until the market-hour it remains on the nest. In old age the upper beak of the eagle grows gradually longer and more crooked, and the bird dies eventually of starvation; there is a folklore story that the eagle is thus punished because it once was a man and refused entertainment to a stranger. The eagle puts aside its superfluous food for its young; for owing to the difficulty in procuring food day by day, it at times may come back to the nest with nothing. If it catch a man prowling about in the neighbourhood of its nest, it will strike him with its wings and scratch him with its talons. The nest is built not on low ground but on an elevated spot, generally on an inaccessible ledge of a cliff; it does, however, build upon a tree. The young are fed until they can fly; hereupon the parent-birds topple them out of the nest, and chase them completely out of the locality. The fact is that a pair of eagles demands an extensive space for its maintenance, and consequently cannot allow other birds to quarter themselves in close neighbourhood. They do not hunt in the vicinity of their nest, but go to a great distance to find their prey. When the eagle has captured a beast, it puts it down without attempting to carry it off at once; if on trial it finds the burden too heavy, it will leave it. When it has spied a hare, it does not swoop on it at once, but lets it go on into the open ground; neither does it descend to the ground at one swoop, but goes gradually down from higher flights to lower and lower: these devices it adopts by way of security against the stratagem of the hunter. It alights on high places by reason of the difficulty it experiences in soaring up from the level ground; it flies high in the air to have the more extensive view; from its high flight it is said to be the only bird that resembles the gods. Birds of prey, as a rule, seldom alight upon rock, as the crookedness of their talons prevents a stable footing on hard stone. The eagle hunts hares, fawns, foxes, and in general all such animals as he can master with ease. It is a long-lived bird, and this fact might be inferred from the length of time during which the same nest is maintained in its place.

Part 33

In Scythia there is found a bird as large as the great bustard. The female lays two eggs, but does not hatch them, but hides them in the skin of a hare or fox and leaves them there, and, when it is not in quest of prey, it keeps a watch on them on a high tree; if any man tries to climb the tree, it fights and strikes him with its wing, just as eagles do.

Part 34

The owl and the night-raven and all the birds see poorly in the daytime seek their prey in the night, but not all the night through, but at evening and dawn. Their food consists of mice, lizards, chafers and the like little creatures. The so-called phene, or lammergeier, is fond of its young, provides its food with ease, fetches food to its nest, and is of a kindly disposition. It rears its own young and those of the eagle as well; for when the eagle ejects its young from the nest, this bird catches them up as they fall and feeds them. For the eagle, by the way, ejects the young birds prematurely, before they are able to feed themselves, or to fly. It appears to do so from jealousy; for it is by nature jealous, and is so ravenous as to grab furiously at its food; and when it does grab at its food, it grabs it in large morsels. It is accordingly jealous of the young birds as they approach maturity, since they are getting good appetites, and so it scratches them with its talons. The young birds fight also with one another, to secure a morsel of food or a comfortable position, whereupon the mother-bird beats them and ejects them from the nest; the young ones scream at this treatment, and the phene hearing them catches them as they fall. The phene has a film over its eyes and sees badly, but the sea-eagle is very keen-sighted, and before its young are fledged tries to make them stare at the sun, and beats the one that refuses to do so, and twists him back in the sun's direction; and if one of them gets watery eyes in the process, it kills him, and rears the other. It lives near the sea, and feeds, as has been said, on sea-birds; when in pursuit of them it catches them one by one, watching the moment when the bird rises to the surface from its dive. When a sea-bird, emerging from the water, sees the sea-eagle, he in terror dives under, intending to rise again elsewhere; the eagle, however, owing to its keenness of vision, keeps flying after him until he either drowns the bird or catches him on the surface. The eagle never attacks these birds when they are in a swarm, for they keep him off by raising a shower of water-drops with their wings.

Part 35

The cepphus is caught by means of sea-foam; the bird snaps at the foam, and consequently fishermen catch it by sluicing with showers of sea-water. These birds grow to be plump and fat;

their flesh has a good odour, excepting the hinder quarters, which smell of shore weed.

Part 36

Of hawks, the strongest is the buzzard; the next in point of courage is the merlin; and the circus ranks third; other diverse kinds are the asterias, the pigeon-hawk, and the pternis; the broaded-winged hawk is called the half-buzzard; others go by the name of hobby-hawk, or sparrow-hawk, or 'smooth-feathered', or 'toad-catcher'. Birds of this latter species find their food with very little difficulty, and flutter along the ground. Some say that there are ten species of hawks, all differing from one another. One hawk, they say, will strike and grab the pigeon as it rests on the ground, but never touch it while it is in flight; another hawk attacks the pigeon when it is perched upon a tree or any elevation, but never touches it when it is on the ground or on the wing; other hawks attack their prey only when it is on the wing. They say that pigeons can distinguish the various species: so that, when a hawk is an assailant, if it be one that attacks its prey when the prey is on the wing, the pigeon will sit still; if it be one that attacks sitting prey, the pigeon will rise up and fly away.

In Thrace, in the district sometimes called that of Cedripolis, men hunt for little birds in the marshes with the aid of hawks. The men with sticks in their hands go beating at the reeds and brushwood to frighten the birds out, and the hawks show themselves overhead and frighten them down. The men then strike them with their sticks and capture them. They give a portion of their booty to the hawks; that is, they throw some of the birds up in the air, and the hawks catch them.

In the neighbourhood of Lake Maeotis, it is said, wolves act in concert with the fishermen, and if the fishermen decline to share with them, they tear their nets in pieces as they lie drying on the shore of the lake.

Part 37

So much for the habits of birds.

In marine creatures, also, one In marine creatures, also, one may observe many ingenious devices adapted to the circumstances of their lives. For the accounts commonly given of the so-called fishing-frog are quite true; as are also those given of the torpedo. The fishing-frog has a set of filaments that project in front of its eyes; they are long and thin like hairs, and are round at the tips; they lie on either side, and are used as baits. Accordingly, when the animal stirs up a place full of sand and mud and conceals itself therein, it raises the filaments, and, when the little fish strike against them, it draws them in underneath into its mouth. The torpedo narcotizes the creatures that it wants to catch, overpowering them by the power of shock that is resident in its body, and feeds upon them; it also hides in the sand and mud, and catches all the creatures that swim in its way and come under its narcotizing influence. This phenomenon has been actually observed in operation. The sting-ray also conceals itself, but not exactly in the same way. That the creatures get their living by this means is obvious from the fact that, whereas they are peculiarly inactive, they are often caught with mullets in their interior, the swiftest of fishes. Furthermore, the fishing-frog is unusually thin when he is caught after losing the tips of his filaments, and the torpedo is known to cause a numbness even in human beings. Again, the hake, the ray, the flat-fish, and the angelfish burrow in the sand, and after concealing themselves angle with the filaments on their mouths, that fishermen call their fishing-rods, and the little creatures on which they feed swim up to the filaments taking them for bits of sea-weed, such as they feed upon.

Wherever an anthias-fish is seen, there will be no dangerous creatures in the vicinity, and sponge-divers will dive in security, and they call these signal-fishes 'holy-fish'. It is a sort of perpetual coincidence, like the fact that wherever snails are present you may be sure there is neither pig nor partridge in the neighbourhood; for both pig and partridge eat up the snails.

The sea-serpent resembles the conger in colour and shape, but is of lesser bulk and more rapid in its movements. If it be caught and thrown away, it will bore a hole with its snout and burrow rapidly in the sand; its snout, by the way, is sharper than that of ordinary serpents. The so-called sea-scolopendra, after swallowing the hook, turns itself inside out until it ejects it, and then it again turns itself outside in. The sea-scolopendra, like the land-scolopendra, will come to a savoury bait; the creature does not bite with its teeth, but stings by contact with its entire body, like the so-called sea-nettle. The so-called fox-shark, when it finds it has swallowed the hook, tries to get rid of it as the scolopendra does, but not in the same way; in other words, it runs up the fishing-line, and bites it off short; it is caught in some districts in deep and rapid waters, with night-lines.

The bonitos swarm together when they espy a dangerous creature, and the largest of them swim round it, and if it touches one of the shoal they try to repel it; they have strong teeth. Amongst other large fish, a lamia-shark, after falling in amongst a shoal, has been seen to be covered with wounds.

Of river-fish, the male of the sheat-fish is remarkably attentive to the young. The female after parturition goes away; the male stays and keeps on guard where the spawn is most abundant, contenting himself with keeping off all other little fishes that might steal the spawn or fry, and this he does for forty or fifty days, until the young are sufficiently grown to make away from the other fishes for themselves. The fishermen can tell where he is on guard: for, in warding off the little fishes, he makes a rush in the water and gives utterance to a kind of muttering noise. He is so earnest in the performance of his parental duties that the fishermen at times, if the eggs be attached to the roots of water-plants deep in the water, drag them into as shallow a place as possible; the male fish will still keep by the young, and, if it so happen, will be caught by the hook when snapping at the little fish that come by; if, however, he be sensible by experience of the danger of the hook, he will still keep by his charge, and with his extremely strong teeth will bite the hook in pieces.

All fish, both those that wander about and those that are stationary, occupy the districts where they were born or very similar places, for their natural food is found there. Carnivorous fish wander most; and all fish are carnivorous with the exception of a few, such as the mullet, the saupe, the red mullet, and the chalcis. The so-called pholis gives out a mucous discharge, which envelops the creature in a kind of nest. Of shell-fish, and fish that are finless, the scallop moves with greatest force and to the greatest distance, impelled along by some internal energy; the murex or purple-fish, and others that resemble it, move hardly at all. Out of the lagoon of Pyrrha all the fishes swim in winter-time, except the sea-gudgeon; they swim out owing to the cold, for the narrow waters are colder than the outer sea, and on the return of the early summer they all swim back again. In the lagoon no scarus is found, nor thritta, nor any other species of the spiny fish, no spotted dogfish, no spiny dogfish, no sea-crawfish, no octopus either of the common or the musky kinds, and certain other fish are also absent; but of fish that are found in the lagoon the white gudgeon is not a marine fish. Of fishes the oviparous are in their prime in the early summer until the spawning time; the viviparous in the autumn, as is also the case with the mullet, the red mullet, and all such fish. In the neighbourhood of Lesbos, the fishes of the outer sea, or of the lagoon, bring forth their eggs or young in the lagoon; sexual union takes place in the autumn, and parturition in the spring. With fishes of the cartilaginous kind, the males and females swarm together in the autumn for the sake of sexual union; in the early summer they come swimming in, and keep apart until after parturition; the two sexes are often taken linked together in sexual union.

Of molluscs the sepia is the most cunning, and is the only species that employs its dark liquid for the sake of concealment as well as from fear: the octopus and calamari make the discharge solely from fear. These creatures never discharge the pigment in its entirety; and after a discharge the pigment accumulates again. The sepia, as has been said, often uses its colouring pigment for concealment; it shows itself in front of the pigment and then retreats back into it; it also hunts with its long tentacles not only little fishes, but oftentimes even mullets. The octopus is a stupid creature, for it will approach a man's hand if it be lowered in the water; but it is neat and thrifty in its habits: that is, it lays up stores in its nest, and, after eating up all that is eatable, it ejects the shells and sheaths of crabs and shell-fish, and the skeletons of little fishes. It seeks its prey by so changing its colour as to render it like the colour of the stones adjacent to it; it does so also when alarmed. By some the sepia is said to perform the same trick; that is, they say it can change its colour so as to make it resemble the colour of its habitat. The only fish that can do this is the angelfish, that is, it can change its colour like the octopus. The octopus as a rule does not live the year out. It has a natural tendency to run off into liquid; for, if beaten and squeezed, it keeps losing substance and at last disappears. The female after parturition is peculiarly subject to this colliquefaction; it becomes stupid; if tossed about by waves, it submits impassively; a man, if he dived, could catch it with the hand; it gets covered over with slime, and makes no effort to catch its wonted prey. The male becomes leathery and clammy. As a proof that they do not live into a second year there is the fact that, after the birth of the little octopuses in the late summer or beginning of autumn, it is seldom that a large-sized octopus is visible, whereas a little before this time of year the creature is at its largest. After the eggs are laid, they say that both the male and the female grow so old and feeble that they are preyed upon by little fish, and with ease dragged from their holes; and that this could not have been done previously; they say also that this is not the case with the small and young octopus, but that the young creature is much stronger than the grown-up one. Neither does the sepia live into a second year. The octopus is the only mollusc that ventures on to dry land; it walks by preference on rough ground; it is firm all over when you squeeze it, excepting in the neck. So much for the mollusca.

It is also said that they make a thin rough shell about them like a hard sheath, and that this is made larger and larger as the animal grows larger, and that it comes out of the sheath as though out of a den or dwelling place.

The nautilus (or argonaut) is a poulpe or octopus, but one peculiar both in its nature and its habits. It rises up from deep water and swims on the surface; it rises with its shell down-turned in order that it may rise the more easily and swim with it empty, but after reaching the surface it shifts the position of the shell. In between its feelers it has a certain amount of web-growth, resembling the substance between the toes of web-footed birds; only that with these latter the substance is thick, while with the nautilus it is thin and like a spider's web. It uses this structure, when a breeze is blowing, for a sail, and lets down some of its feelers alongside as rudder-oars. If it be frightened it fills its shell with water and sinks. With regard to the mode of generation and the growth of the shell knowledge from observation is not yet satisfactory; the shell, however, does not appear to be there from the beginning, but to grow in their cases as in that of other shell-fish; neither is it ascertained for certain whether the animal can live when stripped of the shell.

Part 38

Of all insects, one may also say of all living creatures, the most industrious are the ant, the bee, the hornet, the wasp, and in point of fact all creatures akin to these; of spiders some are more skillful and more resourceful than others. The way in which ants work is open to ordinary observation; how they all march one after the other when they are engaged in putting away and storing up their food; all this may be seen, for they carry on their work even during bright moonlight nights.

Part 39

Of spiders and phalangia there are many species. Of the venomous phalangia there are two; one that resembles the so-called wolf-spider, small, speckled, and tapering to a point; it moves with leaps, from which habit it is nicknamed 'the flea': the other kind is large, black in colour, with long front legs; it is heavy in its movements, walks slowly, is not very strong, and never leaps.

(Of all the other species wherewith poison-vendors supply themselves, some give a weak bite, and others never bite at all. There is another kind, comprising the so-called wolf-spiders.) Of these spiders the small one weaves no web, and the large weaves a rude and poorly built one on the ground or on dry stone walls. It always builds its web over hollow places inside of which it keeps a watch on the end-threads, until some creature gets into the web and begins to struggle, when out the spider pounces. The speckled kind makes a little shabby web under trees.

There is a third species of this animal, preeminently clever and artistic. It first weaves a thread stretching to all the exterior ends of the future web; then from the centre, which it hits upon with great accuracy, it stretches the warp; on the warp it puts what corresponds to the woof, and then weaves the whole together. It sleeps and stores its food away from the centre, but it is at the centre that it keeps watch for its prey. Then, when any creature touches the web and the centre is set in motion, it first ties and wraps the creature round with threads until it renders it helpless, then lifts it and carries it off, and, if it happens to be hungry, sucks out the life-juices--for that is the way it feeds; but, if it be not hungry, it first mends any damage done and then hastens again to its quest of prey. If something comes meanwhile into the net, the spider at first makes for the centre, and then goes back to its entangled prey as from a fixed starting point. If any one injures a portion of the web, it recommences weaving at sunrise or at sunset, because it is chiefly at these periods that creatures are caught in the web. It is the female that does the weaving and the hunting, but the male takes a share of the booty captured.

Of the skilful spiders, weaving a substantial web, there are two kinds, the larger and the smaller. The one has long legs and keeps watch while swinging downwards from the web: from its large size it cannot easily conceal itself, and so it keeps underneath, so that its prey may not be frightened off, but may strike upon the web's upper surface; the less awkwardly formed one lies in wait on the top, using a little hole for a lurking-place. Spiders can spin webs from the time of their birth, not from their interior as a superfluity or excretion, as Democritus avers, but off their body as a kind of tree-bark, like the creatures that shoot out with their hair, as for instance the porcupine. The creature can attack animals larger than itself, and enwrap them with its threads: in other words, it will attack a small lizard, run round and draw threads about its mouth until it closes the mouth up; then it comes up and bites it.

Part 40

So much for the spider. Of insects, there is a genus that has no one name that comprehends all the species, though all the species are akin to one another in form; it consists of all the insects that construct a honeycomb: to wit, the bee, and all the insects that resemble it in form.

There are nine varieties, of which six are gregarious-the bee, the king-bee, the drone bee, the annual wasp, and, furthermore, the anthrene (or hornet), and the tenthredo (or ground-wasp); three are solitary-the smaller siren, of a dun colour, the larger siren, black and speckled, and the third, the largest of all, that is called the humble-bee. Now ants never go a-hunting, but gather up what is ready to hand; the spider makes nothing, and lays up no store, but simply goes a-hunting for its food; while the bee--for we shall by and by treat of the nine varieties--does not go a-hunting, but constructs its food out of gathered material and stores it away, for honey is the bee's food. This fact is shown by the beekeepers' attempt to remove the combs; for the bees, when they are fumigated, and are suffering great distress from the process, then devour the honey most ravenously, whereas at other times they are never observed to be so greedy, but apparently are thrifty and disposed to lay by for their future sustenance. They have also another food which is called bee-bread; this is scarcer than honey and has a sweet figlike taste; this they carry as they do the wax on their legs.

Very remarkable diversity is observed in their methods of working and their general habits. When the hive has been delivered to them clean and empty, they build their waxen cells, bringing in the juice of all kinds of flowers and the 'tears' or exuding sap of trees, such as willows and elms and such others as are particularly given to the exudation of gum. With this material they besmear the groundwork, to provide against attacks of other creatures; the bee-keepers call this stuff 'stop-wax'. They also with the same material narrow by side-building the entrances to the hive if they are too wide. They first build cells for themselves; then for the so-called kings and the drones; for themselves they are always building, for the kings only when the brood of young is numerous, and cells for the drones they build if a superabundance of honey should suggest their doing so. They build the royal cells next to their own, and they are of small bulk; the drones' cells they build nearby, and these latter are less in bulk than the bee's cells.

They begin building the combs downwards from the top of the hive, and go down and down building many combs connected together until they reach the bottom. The cells, both those for the honey and those also for the grubs, are double-doored; for two cells are ranged about a single base, one pointing one way and one the other, after the manner of a double (or hour-glass-shaped) goblet. The cells that lie at the commencement of the combs and are attached to the hives, to the extent of two or three concentric circular rows, are small and devoid of honey; the cells that are well filled with honey are most thoroughly luted with wax. At the entry to the hive the aperture of the doorway is smeared with mitys; this substance is a deep black, and is a sort of dross or residual by-product of wax; it has a pungent odour, and is a cure for bruises and suppurating sores. The greasy stuff that comes next is pitch-wax; it has a less pungent odour and is less medicinal than the mitys. Some say that the drones construct combs by themselves in the same hive and in the same comb that they share with the bees; but that they make no honey, but subsist, they and their grubs also, on the honey made by the bees. The drones, as a rule, keep inside the hive; when they go out of doors, they soar up in the air in a stream, whirling round and round in a kind of gymnastic exercise; when this is over, they come inside the hive and feed to repletion ravenously. The kings never quit the hive, except in conjunction with the entire swarm, either for food or for any other reason. They say that, if a young swarm go astray, it will turn back upon its route and by the aid of scent seek out its leader. It is said that if he is unable to fly he is carried by the swarm, and that if he dies the swarm perishes; and that, if this swarm outlives the king for a while and constructs combs, no honey is produced and the bees soon die out.

Bees scramble up the stalks of flowers and rapidly gather the bees-wax with their front legs; the front legs wipe it off on to the middle legs, and these pass it on to the hollow curves of the hind-legs; when thus laden, they fly away home, and one may see plainly that their load is a heavy one. On each expedition the bee does not fly from a flower of one kind to a flower of another, but flies from one violet, say, to another violet, and never meddles with another flower until it has got back to the hive; on reaching the hive they throw off their load, and each bee on his return is accompanied by three or four companions. One cannot well tell what is the substance they gather, nor the exact process of their work. Their mode of gathering wax has been observed on olive-trees, as owing to the thickness of the leaves the bees remain stationary for a considerable while. After this work is over, they attend to the grubs. There is nothing to prevent grubs, honey, and drones being all found in one and the same comb. As long as the leader is alive, the drones are said to be produced apart by themselves; if he be no longer living, they are said to be reared by the bees in their own cells, and under these circumstances to become more spirited: for this reason they are called 'sting-drones', not that they really have stings, but that they have the wish without the power, to use such weapons. The cells for the drones are larger than the others; sometimes the bees construct cells for the drones apart, but usually they put them in amongst their own; and when this is the case the bee-keepers cut the drone-cells out of the combs.

There are several species of bees, as has been said; two of 'kings', the better kind red, the other black and variegated, and twice as big as the working-bee. The best workingbee is small, round, and speckled: another kind is long and like an anthrene wasp; another kind is what is called the robber-bee, black and flat-bellied; then there is the drone, the largest of all, but devoid of sting, and lazy. There is a difference between the progeny of bees that inhabit cultivated land and of those from the mountains: the forest-bees are more shaggy, smaller, more industrious and more fierce. Working-bees make their combs all even, with the superficial covering quite smooth. Each comb is of one kind only: that is, it contains either bees only, or grubs only, or drones only; if it happen, however, that they make in one and the same comb all these kinds of cells, each separate kind will be built in a continuous row right through. The long bees build uneven combs, with the lids of the cells protuberant, like those of the anthrene; grubs and everything else have no fixed places, but are put anywhere; from these bees come inferior kings, a large quantity of drones, and the so-called robber-bee; they produce either no honey at all, or honey in very small quantities. Bees brood over the combs and so mature them; if they fail to do so, the combs are said to go bad and to get covered with a sort of spider's web. If they can keep brooding over the part undamaged, the damaged part simply eats itself away; if they cannot so brood, the entire comb perishes; in the damaged combs small worms are engendered, which take on wings and fly away. When the combs keep settling down, the bees restore the level surface, and put props underneath the combs to give themselves free passage-room; for if such free passage be lacking they cannot brood, and the cobwebs come on. When the robber-bee and the drone appear, not only do they do no work themselves, but they actually damage the work of the other bees; if they are caught in the act, they are killed by the working-bees. These bees also kill without mercy most of their kings, and especially kings of the inferior sort; and this they do for fear a multiplicity of kings should lead to a dismemberment of the hive. They kill them especially when the hive is deficient in grubs, and a swarm is not intended to take place; under these circumstances they destroy the cells of the kings if they have been prepared, on the ground that these kings are always ready to lead out swarms. They destroy also the combs of the drones if a failure in the supply be threatening and the hive runs short of provisions; under such circumstances they fight desperately with all who try to take their honey, and eject from the hive all the resident drones; and oftentimes the drones are to be seen sitting apart in the hive. The little bees fight vigorously with the long kind, and try to banish them from the hives; if they succeed, the hive will be unusually productive, but if the bigger bees get left mistresses of the field they pass the time in idleness, and no good at all but die out before the autumn. Whenever the working-bees kill an enemy they try to do so out of doors; and whenever one of their own body dies, they carry the dead bee out of doors also. The so-called robber-bees spoil their own combs, and, if they can do so unnoticed, enter and spoil the combs of other bees; if they are caught in the act they are put to death. It is no easy task for them to escape detection, for there are sentinels on guard at every entry; and, even if they do escape detection on entering, afterwards from a surfeit of food they cannot fly, but go rolling about in front of the hive, so that their chances of escape are small indeed. The kings are never themselves seen outside the hive except with a swarm in flight: during which time all the other bees cluster around them. When the flight of a swarm is imminent, a monotonous and quite peculiar sound made by all the bees is heard for several days, and for two or three days in advance a few bees are seen flying round the hive; it has never as yet been ascertained, owing to the difficulty of the observation, whether or no the king is among these. When they have swarmed, they fly away and separate off to each of the kings; if a small swarm happens to settle near to a large one, it will shift to join this large one, and if the king whom they have abandoned follows them, they put him to death. So much for the quitting of the hive and the swarmflight. Separate detachments of bees are told off for diverse operations; that is, some carry flower-produce, others carry water, others smooth and arrange the combs. A bee carries water when it is rearing grubs. No bee ever settles on the flesh of any creature, or ever eats animal food. They have no fixed date for commencing work; but when their provender is forthcoming and they are in comfortable trim, and by preference in summer, they set to work, and when the weather is fine they work incessantly.

The bee, when quite young and in fact only three days old, after shedding its chrysalis-case, begins to work if it be well fed. When a swarm is settling, some bees detach themselves in search of food and return back to the swarm. In hives that are in good condition the production of young bees is discontinued only for the forty days that follow the winter solstice. When the grubs are grown, the bees put food beside them and cover them with a coating of wax; and, as soon as the grub is strong enough, he of his own accord breaks the lid and comes out. Creatures that make their appearance in hives and spoil the combs the working-bees clear out, but the other bees from sheer laziness look with indifference on damage done to their produce. When the bee-masters take out the combs, they leave enough food behind for winter use; if it be sufficient in quantity, the occupants of the hive will survive; if it be insufficient, then, if the weather be rough, they die on the spot, but if it be fair, they fly away and desert the hive. They feed on honey summer and winter; but they store up another article of food resembling wax in hardness, which by some is called sandarace, or bee-bread. Their worst enemies are wasps and the birds named titmice, and furthermore the swallow and the bee-eater. The frogs in the marsh also catch them if they come in their way by the water-side, and for this reason bee-keepers chase the frogs from the ponds from which the bees take water; they destroy also wasps' nests, and the nests of swallows, in the neighbourhood of the hives, and also the nests of bee-eaters. Bees have fear only of one another. They fight with one another and with wasps. Away from the hive they attack neither their own species nor any other creature, but in the close proximity of the hive they kill whatever they get hold of. Bees that sting die from their inability to extract the sting without at the same time extracting their intestines. True, they often recover, if the person stung takes the trouble to press the sting out; but once it loses its sting the bee must die. They can kill with their stings even large animals; in fact, a horse has been known to have been stung to death by them. The kings are the least disposed to show anger or to inflict a sting. Bees that die are removed from the hive, and in every way the creature is remarkable for its cleanly habits; in point of fact, they often fly away to a distance to void their excrement because it is malodorous; and, as has been said, they are annoyed by all bad smells and by the scent of perfumes, so much so that they sting people that use perfumes.

They perish from a number of accidental causes, and when their kings become too numerous and try each to carry away a portion of the swarm.

The toad also feeds on bees; he comes to the doorway of the hive, puffs himself out as he sits on the watch, and devours the creatures as they come flying out; the bees can in no way retaliate, but the bee-keeper makes a point of killing him.

As for the class of bee that has been spoken of as inferior or good-for-nothing, and as constructing its combs so roughly, some bee-keepers say that it is the young bees that act so from inexperience; and the bees of the current year are termed young. The young bees do not sting as the others do; and it is for this reason that swarms may be safely carried, as it is of young bees that they are composed. When honey runs short they expel the drones, and the bee-keepers supply the bees with figs and sweet-tasting articles of food. The elder bees do the indoor work, and are rough and hairy from staying indoors; the young bees do the outer carrying, and are comparatively smooth. They kill the drones also when in their work they are confined for room; the drones, by the way, live in the innermost recess of the hive. On one occasion, when a hive was in a poor condition, some of the occupants assailed a foreign hive; proving victorious in a combat they took to carrying off the honey; when the bee-keeper tried to kill them, the other bees came out and tried to beat off the enemy but made no attempt to sting the man.

The diseases that chiefly attack prosperous hives are first of all the clerus-this consists in a growth of little worms on the floor, from which, as they develop, a kind of cobweb grows over the entire hive, and the combs decay; another diseased condition is indicated in a lassitude on the part of the bees and in malodorousness of the hive. Bees feed on thyme; and the white thyme is better than the red. In summer the place for the hive should be cool, and in winter warm. They are very apt to fall sick if the plant they are at work on be mildewed. In a high wind they carry a stone by way of ballast to steady them. If a stream be near at hand, they drink from it and from it only, but before they drink they first deposit their load; if there be no water near at hand, they disgorge their honey as they drink elsewhere, and at once make off to work. There are two seasons for making honey, spring and autumn; the spring honey is sweeter, whiter, and in every way better than the autumn honey. Superior honey comes from fresh comb, and from young shoots; the red honey is inferior, and owes its inferiority to the comb in which it is deposited, just as wine is apt to be spoiled by its cask; consequently, one should have it looked to and dried. When the thyme is in flower and the comb is full, the honey does not harden. The honey that is golden in hue is excellent. White honey does not come from thyme pure and simple; it is good as a salve for sore eyes and wounds. Poor honey always floats on the surface and should be skimmed off; the fine clear honey rests below. When the floral world is in full bloom, then they make wax; consequently you must then take the wax out of the hive, for they go to work on new wax at once. The flowers from which they gather honey are as follows: the spindle-tree, the melilot-clover, king's-spear, myrtle, flowering-reed, withy, and broom. When they work at thyme, they mix in water before sealing up the comb. As has been already stated, they all either fly to a distance to discharge their excrement or make the discharge into one single comb. The little bees, as has been said, are more industrious than the big ones; their wings are battered; their colour is black, and they have a burnt-up aspect. Gaudy and showy bees, like gaudy and showy women, are good-for-nothings.

Bees seem to take a pleasure in listening to a rattling noise; and consequently men say that they can muster them into a hive by rattling with crockery or stones; it is uncertain, however, whether or no they can hear the noise at all and also whether their procedure is due to pleasure or alarm. They expel from the hive all idlers and unthrifts. As has been said, they differentiate their work; some make wax, some make honey, some make bee-bread, some shape and mould combs, some bring water to the cells and mingle it with the honey, some engage in out-of-door work. At early dawn they make no noise, until some one particular bee makes a buzzing noise two or three times and thereby awakes the rest; hereupon they all fly in a body to work. By and by they return and at first are noisy; then the noise gradually decreases, until at last some one bee flies round about, making a buzzing noise, and apparently calling on the others to go to sleep; then all of a sudden there is a dead silence.

The hive is known to be in good condition if the noise heard within it is loud, and if the bees make a flutter as they go out and in; for at this time they are constructing brood-cells. They suffer most from hunger when they recommence work after winter. They become somewhat lazy if the bee-keeper, in robbing the hive, leave behind too much honey; still one should leave cells numerous in proportion to the population, for the bees work in a spiritless way if too few combs are left. They become idle also, as being dispirited, if the hive be too big. A hive yields to the bee-keeper six or nine pints of honey; a prosperous hive will yield twelve or fifteen pints, exceptionally good hives eighteen. Sheep and, as has been said, wasps are enemies to the bees. Bee-keepers entrap the latter, by putting a flat dish on the ground with pieces of meat on it; when a number of the wasps settle on it, they cover them with a lid and put the dish and its contents on the fire. It is a good thing to have a few drones in a hive, as their presence increases the industry of the workers. Bees can tell the approach of rough weather or of rain; and the proof is that they will not fly away, but even while it is as yet fine they go fluttering about within a restricted space, and the bee-keeper knows from this that they are expecting bad weather. When the bees inside the hive hang clustering to one another, it is a sign that the swarm is intending to quit; consequently, occasion, when a bee-keepers, on seeing this, besprinkle the hive with sweet wine. It is advisable to plant about the hives pear-trees, beans, Median-grass, Syrian-grass, yellow pulse, myrtle, poppies, creeping-thyme, and almond-trees. Some bee-keepers sprinkle their bees with flour, and can distinguish them from others when they are at work out of doors. If the spring be late, or if there be drought or blight, then grubs are all the fewer in the hives. So much for the habits of bees.

Part 41

Of wasps, there are two kinds. Of these kinds one is wild and scarce, lives on the mountains, engenders grubs not underground but on oak-trees, is larger, longer, and blacker than the other kind, is invariably speckled and furnished with a sting, and is remarkably courageous. The pain from its sting is more severe than that caused by the others, for the instrument that causes the pain is larger, in proportion to its own larger size. These wild live over into a second year, and in winter time, when oaks have been in course of felling, they may be seen coming out and flying away. They lie concealed during the winter, and live in the interior of logs of wood. Some of them are mother-wasps and some are workers, as with the tamer kind; but it is by observation of the tame wasps that one may learn the varied characteristics of the mothers and the workers. For in the case of the tame wasps also there are two kinds; one consists of leaders, who are called mothers, and the other of workers. The leaders are far larger and milder-tempered than the others. The workers do not live over into a second year, but all die when winter comes on; and this can be proved, for at the commencement of winter the workers become drowsy, and about the time of the winter solstice they are never seen at all. The leaders, the so-called mothers, are seen all through the winter, and live in holes underground; for men when ploughing or digging in winter have often come upon mother-wasps, but never upon workers. The mode of reproduction of wasps is as follows. At the approach of summer, when the leaders have found a sheltered spot, they take to moulding their combs, and construct the so-called sphecons,-little nests containing four cells or thereabouts, and in these are produced working-wasps but not mothers. When these are grown up, then they construct other larger combs upon the first, and then again in like manner others; so that by the close of autumn there are numerous large combs in which the leader, the so-called mother, engenders no longer working-wasps but mothers. These develop high up in the nest as large grubs, in cells that occur in groups of four or rather more, pretty much in the same way as we have seen the grubs of the king-bees to be produced in their cells. After the birth of the working-grubs in the cells, the leaders do nothing and the workers have to supply them with nourishment; and this is inferred from the fact that the leaders (of the working-wasps) no longer fly out at this time, but rest quietly indoors. Whether the leaders of last year after engendering new leaders are killed by the new brood, and whether this occurs invariably or whether they can live for a longer time, has not been ascertained by actual observation; neither can we speak with certainty, as from observation, as to the age attained by the mother-wasp or by the wild wasps, or as to any other similar phenomenon. The mother-wasp is broad and heavy, fatter and larger than the ordinary wasp, and from its weight not very strong on the wing; these wasps cannot fly far, and for this reason they always rest inside the nest, building and managing its indoor arrangements. The so-called mother-wasps are found in most of the nests; it is a matter of doubt whether or no they are provided with stings; in all probability, like the king-bees, they have stings, but never protrude them for offence. Of the ordinary wasps some are destitute of stings, like the drone-bees, and some are provided with them. Those unprovided therewith are smaller and less spirited and never fight, while the others are big and courageous; and these latter, by some, are called males, and the stingless, females. At the approach of winter many of the wasps that have stings appear to lose them; but we have never met an eyewitness of this phenomenon. Wasps are more abundant in times of drought and in wild localities. They live underground; their combs they mould out of chips and earth, each comb from a single origin, like a kind of root. They feed on certain flowers and fruits, but for the most part on animal food. Some of the tame wasps have been observed when sexually united, but it was not determined whether both, or neither, had stings, or whether one had a sting and the other had not; wild wasps have been seen under similar circumstances, when one was seen to have a sting but the case of the other was left undetermined. The wasp-grub does not appear to come into existence by parturition, for at the outset the grub is too big to be the offspring of a wasp. If you take a wasp by the feet and let him buzz with the vibration of his wings, wasps that have no stings will fly toward it, and wasps that have stings will not; from which fact it is inferred by some that one set are males and the other females. In holes in the ground in winter-time wasps are found, some with stings, and some without. Some build cells, small and few in number; others build many and large ones. The so-called mothers are caught at the change of season, mostly on elm-trees, while gathering a substance sticky and gumlike. A large number of mother-wasps are found when in the previous year wasps have been numerous and the weather rainy; they are captured in precipitous places, or in vertical clefts in the ground, and they all appear to be furnished with stings.

Part 42

So much for the habits of wasps. Anthrenae do not subsist by culling from flowers as bees do, but for the most part on animal food: for this reason they hover about dung; for they chase the large flies, and after catching them lop off their heads and fly away with the rest of the carcases; they are furthermore fond of sweet fruits. Such is their food. They have also kings or leaders like bees and wasps; and their leaders are larger in proportion to themselves than are wasp-kings to wasps or bee-kings to bees. The anthrena-king, like the wasp-king, lives indoors. Anthrenae build their nests underground, scraping out the soil like ants; for neither anthrenae nor wasps go off in swarms as bees do, but successive layers of young anthrenae keep to the same habitat, and go on enlarging their nest by scraping out more and more of soil. The nest accordingly attains a great size; in fact, from a particularly prosperous nest have been removed three and even four baskets full of combs. They do not, like bees, store up food, but pass the winter in a torpid condition; the greater part of them die in the winter, but it is uncertain whether that can be said of them all, In the hives of bees several kings are found and they lead off detachments in swarms, but in the anthrena's nest only one king is found. When individual anthrenae have strayed from their nest, they cluster on a tree and construct combs, as may be often seen above-ground, and in this nest they produce a king; when the king is full-grown, he leads them away and settles them along with himself in a hive or nest. With regard to their sexual unions, and the method of their reproduction, nothing is known from actual observation. Among bees both the drones and the kings are stingless, and so are certain wasps, as has been said; but anthrenae appear to be all furnished with stings: though, by the way, it would well be worthwhile to carry out investigation as to whether the anthrena-king has a sting or not.

Part 43

Humble-bees produce their young under a stone, right on the ground, in a couple of cells or little more; in these cells is found an attempt at honey, of a poor description. The tenthredon is like the anthrena, but speckled, and about as broad as a bee. Being epicures as to their food, they fly, one at a time, into kitchens and on to slices of fish and the like dainties. The tenthredon brings forth, like the wasp, underground, and is very prolific; its nest is much bigger and longer than that of the wasp. So much for the methods of working and the habits of life of the bee, the wasp, and all the other similar insects.

Part 44

As regards the disposition or temper of animals, as has been previously observed, one may detect great differences in respect to courage and timidity, as also, even among wild animals, in regard to tameness and wildness. The lion, while he is eating, is most ferocious; but when he is not hungry and has had a good meal, he is quite gentle. He is totally devoid of suspicion or nervous fear, is fond of romping with animals that have been reared along with him and to whom he is accustomed, and manifests great affection towards them. In the chase, as long as he is in view, he makes no attempt to run and shows no fear, but even if he be compelled by the multitude of the hunters to retreat, he withdraws deliberately, step by step, every now and then turning his head to regard his pursuers. If, however, he reach wooded cover, then he runs at full speed, until he comes to open ground, when he resumes his leisurely retreat. When, in the open, he is forced by the number of the hunters to run while in full view, he does run at the top of his speed, but without leaping and bounding. This running of his is evenly and continuously kept up like the running of a dog; but when he is in pursuit of his prey and is close behind, he makes a sudden pounce upon it. The two statements made regarding him are quite true; the one that he is especially afraid of fire, as Homer pictures him in the line-'and glowing torches, which, though fierce he dreads,'-and the other, that he keeps a steady eye upon the hunter who hits him, and flings himself upon him. If a hunter hit him, without hurting him, then if with a bound he gets hold of him, he will do him no harm, not even with his claws, but after shaking him and giving him a fright will let him go again. They invade the cattle-folds and attack human beings when they are grown old and so by reason of old age and the diseased condition of their teeth are unable to pursue their wonted prey. They live to a good old age. The lion who was captured when lame, had a number of his teeth broken; which fact was regarded by some as a proof of the longevity of lions, as he could hardly have been reduced to this condition except at an advanced age. There are two species of lions, the plump, curly-maned, and the long-bodied, straight maned; the latter kind is courageous, and the former comparatively timid; sometimes they run away with their tail between their legs, like a dog. A lion was once seen to be on the point of attacking a boar, but to run away when the boar stiffened his bristles in defence. It is susceptible of hurt from a wound in the flank, but on any other part of its frame will endure any number of blows, and its head is especially hard. Whenever it inflicts a wound, either by its teeth or its claws, there flows from the wounded parts suppurating matter, quite yellow, and not to be stanched by bandage or sponge; the treatment for such a wound is the same as that for the bite of a dog.

The thos, or civet, is fond of man's company; it does him no harm and is not much afraid of him, but it is an enemy to the dog and the lion, and consequently is not found in the same habitat with them. The little ones are the best. Some say that there are two species of the animal, and some say, three; there are probably not more than three, but, as is the case with certain of the fishes, birds, and quadrupeds, this animal changes in appearance with the change of season. His colour in winter is not the same as it is in summer; in summer the animal is smooth-haired, in winter he is clothed in fur.

Part 45

The bison is found in Paeonia on Mount Messapium, which separates Paeonia from Maedica; and the Paeonians call it the monapos. It is the size of a bull, but stouter in build, and not long in the body; its skin, stretched tight on a frame, would give sitting room for seven people. In general it resembles the ox in appearance, except that it has a mane that reaches down to the point of the shoulder, as that of the horse reaches down to its withers; but the hair in its mane is softer than the hair in the horse's mane, and clings more closely. The colour of the hair is brown-yellow; the mane reaches down to the eyes, and is deep and thick. The colour of the body is half red, half ashen-grey, like that of the so-called chestnut horse, but rougher. It has an undercoat of woolly hair. The animal is not found either very black or very red. It has the bellow of a bull. Its horns are crooked, turned inwards towards each other and useless for purposes of self-defence; they are a span broad, or a little more, and in volume each horn would hold about three pints of liquid; the black colour of the horn is beautiful and bright. The tuft of hair on the forehead reaches down to the eyes, so that the animal sees objects on either flank better than objects right in front. It has no upper teeth, as is the case also with kine and all other horned animals. Its legs are hairy; it is cloven-footed, and the tail, which resembles that of the ox, seems not big enough for the size of its body. It tosses up dust and scoops out the ground with its hooves, like the bull. Its skin is impervious to blows. Owing to the savour of its flesh it is sought for in the chase. When it is wounded it runs away, and stops only when thoroughly exhausted. It defends itself against an assailant by kicking and projecting its excrement to a distance of eight yards; this device it can easily adopt over and over again, and the excrement is so pungent that the hair of hunting-dogs is burnt off by it. It is only when the animal is disturbed or alarmed that the dung has this property; when the animal is undisturbed it has no blistering effect. So much for the shape and habits of the animal. When the season comes for parturition the mothers give birth to their young in troops upon the mountains. Before dropping their young they scatter their dung in all directions, making a kind of circular rampart around them; for the animal has the faculty of ejecting excrement in most extraordinary quantities.

Part 46

Of all wild animals the most easily tamed and the gentlest is the elephant. It can be taught a number of tricks, the drift and meaning of which it understands; as, for instance, it can taught to kneel in presence of the king. It is very sensitive, and possessed of an intelligence superior to that of other animals. When the male has had sexual union with the female, and the female has conceived, the male has no further intercourse with her.

Some say that the elephant lives for two hundred years; others, for one hundred and twenty; that the female lives nearly as long as the male; that they reach their prime about the age of sixty, and that they are sensitive to inclement weather and frost. The elephant is found by the banks of rivers, but he is not a river animal; he can make his way through water, as long as the tip of his trunk can be above the surface, for he blows with his trunk and breathes through it. The animal is a poor swimmer owing to the heavy weight of his body.

Part 47

The male camel declines intercourse with its mother; if his keeper tries compulsion, he evinces disinclination. On one occasion, when intercourse was being declined by the young male, the keeper covered over the mother and put the young male to her; but, when after the intercourse the wrapping had been removed, though the operation was completed and could not be revoked, still by and by he bit his keeper to death. A story goes that the king of Scythia had a highly-bred mare, and that all her foals were splendid; that wishing to mate the best of the young males with the mother, he had him brought to the stall for the purpose; that the young horse declined; that, after the mother's head had been concealed in a wrapper he, in ignorance, had intercourse; and that, when immediately afterwards the wrapper was removed and the head of the mare was rendered visible, the young horse ran way and hurled himself down a precipice.

Part 48

Among the sea-fishes many stories are told about the dolphin, indicative of his gentle and kindly nature, and of manifestations of passionate attachment to boys, in and about Tarentum, Caria, and other places. The story goes that, after a dolphin had been caught and wounded off the coast of Caria, a shoal of dolphins came into the harbour and stopped there until the fisherman let his captive go free; whereupon the shoal departed. A shoal of young dolphins is always, by way of protection, followed by a large one. On one occasion a shoal of dolphins, large and small, was seen, and two dolphins at a little distance appeared swimming in underneath a little dead dolphin when it was sinking, and supporting it on their backs, trying out of compassion to prevent its being devoured by some predaceous fish. Incredible stories are told regarding the rapidity of movement of this creature. It appears to be the fleetest of all animals, marine and terrestrial, and it can leap over the masts of large vessels. This speed is chiefly manifested when they are pursuing a fish for food; then, if the fish endeavours to escape, they pursue him in their ravenous hunger down to deep waters; but, when the necessary return swim is getting too long, they hold in their breath, as though calculating the length of it, and then draw themselves together for an effort and shoot up like arrows, trying to make the long ascent rapidly in order to breathe, and in the effort they spring right over the a ship's masts if a ship be in the vicinity. This same phenomenon is observed in divers, when they have plunged into deep water; that is, they pull themselves together and rise with a speed proportional to their strength. Dolphins live together in pairs, male and female. It is not known for what reason they run themselves aground on dry land; at all events, it is said that they do so at times, and for no obvious reason.

Part 49

Just as with all animals a change of action follows a change of circumstance, so also a change of character follows a change of action, and often some portions of the physical frame undergo a change, occurs in the case of birds. Hens, for instance, when they have beaten the cock in a fight, will crow like the cock and endeavour to tread him; the crest rises up on their head and the tail-feathers on the rump, so that it becomes difficult to recognize that they are hens; in some cases there is a growth of small spurs. On the death of a hen a cock has been seen to undertake the maternal duties, leading the chickens about and providing them with food, and so intent upon these duties as to cease crowing and indulging his sexual propensities. Some cock-birds are congenitally so feminine that they will submit patiently to other males who attempt to tread them.

Part 50

Some animals change their form and character, not only at certain ages and at certain seasons, but in consequence of being castrated; and all animals possessed of testicles may be submitted to this operation. Birds have their testicles inside, and oviparous quadrupeds close to the loins; and of viviparous animals that walk some have them inside, and most have them outside, but all have them at the lower end of the belly. Birds are castrated at the rump at the part where the two sexes unite in copulation. If you burn this twice or thrice with hot irons, then, if the bird be full-grown, his crest grows sallow, he ceases to crow, and foregoes sexual passion; but if you cauterize the bird when young, none of these male attributes propensities will come to him as he grows up. The case is the same with men: if you mutilate them in boyhood, the later-growing hair never comes, and the voice never changes but remains high-pitched; if they be mutilated in early manhood, the late growths of hair quit them except the growth on the groin, and that diminishes but does not entirely depart. The congenital growths of hair never fall out, for a eunuch never grows bald. In the case of all castrated or mutilated male quadrupeds the voice changes to the feminine voice. All other quadrupeds when castrated, unless the operation be performed when they are young, invariably die; but in the case of boars, and in their case only, the age at which the operation is performed produces no difference. All animals, if operated on when they are young, become bigger and better looking than their unmutilated fellows; if they be mutilated when full-grown, they do not take on any increase of size. If stags be mutilated, when, by reason of their age, they have as yet no horns, they never grow horns at all; if they be mutilated when they have horns, the horns remain unchanged in size, and the animal does not lose them. Calves are mutilated when a year old; otherwise, they turn out uglier and smaller. Steers are mutilated in the following way: they turn the animal over on its back, cut a little off the scrotum at the lower end, and squeeze out the testicles, then push back the roots of them as far as they can, and stop up the incision with hair to give an outlet to suppurating matter; if inflammation ensues, they cauterize the scrotum and put on a plaster. If a full-grown bull be mutilated, he can still to all appearance unite sexually with the cow. The ovaries of sows are excised with the view of quenching in them sexual appetites and of stimulating growth in size and fatness. The sow has first to be kept two days without food, and, after being hung up by the hind legs, it is operated on; they cut the lower belly, about the place where the boars have their testicles, for it is there that the ovary grows, adhering to the two divisions (or horns) of the womb; they cut off a little piece and stitch up the incision. Female camels are mutilated when they are wanted for war purposes, and are mutilated to prevent their being got with young. Some of the inhabitants of Upper Asia have as many as three thousand camels: when they run, they run, in consequence of the length of their stride, much quicker than the horses of Nisaea. As a general rule, mutilated animals grow to a greater length than the unmutilated.

All animals that ruminate derive profit and pleasure from the process of rumination, as they do from the process of eating. It is the animals that lack the upper teeth that ruminate, such as kine, sheep, and goats. In the case of wild animals no observation has been possible; save in the case of animals that are occasionally domesticated, such as the stag, and it, we know, chews the cud. All animals that ruminate generally do so when lying down on the ground. They carry on the process to the greatest extent in winter, and stall-fed ruminants carry it on for about seven months in the year; beasts that go in herds, as they get their food out of doors, ruminate to a lesser degree and over a lesser period. Some, also, of the animals that have teeth in both jaws ruminate; as, for instance, the Pontic mice, and the fish which from the habit is by some called 'the Ruminant', (as well as other fish).

Long-limbed animals have loose feces, and broad-chested animals vomit with comparative facility, and these remarks are, in a general way, applicable to quadrupeds, birds, and men.

Part 51

A considerable number of birds change according to season the colour of their plumage and their note; as, for instance, the owsel becomes yellow instead of black, and its note gets altered, for in summer it has a musical note and in winter a discordant chatter. The thrush also changes its colour; about the throat it is marked in winter with speckles like a starling, in summer distinctly spotted: however, it never alters its note. The nightingale, when the hills are taking on verdure, sings continually for fifteen days and fifteen nights; afterwards it sings, but not continuously. As summer advances it has a different song, not so varied as before, nor so deep, nor so intricately modulated, but simple; it also changes its colour, and in Italy about this season it goes by a different name. It goes into hiding, and is consequently visible only for a brief period.

The erithacus (or redbreast) and the so-called redstart change into one another; the former is a winter bird, the latter a summer one, and the difference between them is practically limited to the coloration of their plumage. In the same way with the beccafico and the blackcap; these change into one another. The beccafico appears about autumn, and the blackcap as soon as autumn has ended. These birds, also, differ from one another only in colour and note; that these birds, two in name, are one in reality is proved by the fact that at the period when the change is in progress each one has been seen with the change as yet incomplete. It is not so very strange that in these cases there is a change in note and in plumage, for even the ring-dove ceases to coo in winter, and recommences cooing when spring comes in; in winter, however, when fine weather has succeeded to very stormy weather, this bird has been known to give its cooing note, to the astonishment of such as were acquainted with its usual winter silence. As a general rule, birds sing most loudly and most diversely in the pairing season. The cuckoo changes its colour, and its note is not clearly heard for a short time previous to its departure. It departs about the rising of the Dog-star, and it reappears from springtime to the rising of the Dog-star. At the rise of this star the bird called by some oenanthe disappears, and reappears when it is setting: thus keeping clear at one time of extreme cold, and at another time of extreme heat. The hoopoe also changes its colour and appearance, as Aeschylus has represented in the following lines:-

The Hoopoe, witness to his own distress,

Is clad by Zeus in variable dress:-

Now a gay mountain-bird, with knightly crest,

Now in the white hawk's silver plumage drest,

For, timely changing, on the hawk's white wing

He greets the apparition of the Spring.

Thus twofold form and colour are conferred,

In youth and age, upon the selfsame bird.

The spangled raiment marks his youthful days,

The argent his maturity displays;

And when the fields are yellow with ripe corn

Again his particoloured plumes are worn.

But evermore, in sullen discontent,

He seeks the lonely hills, in self-sought banishment.

Of birds, some take a dust-bath by rolling in dust, some take a water-bath, and some take neither the one bath nor the other. Birds that do not fly but keep on the ground take the dust-bath, as for instance the hen, the partridge, the francolin, the crested lark, the pheasant; some of the straight-taloned birds, and such as live on the banks of a river, in marshes, or by the sea, take a water-bath; some birds take both the dust-bath and the waterbath, as for instance the pigeon and the sparrow; of the crooked-taloned birds the greater part take neither the one bath nor the other. So much for the ways of the above-mentioned, but some birds have a peculiar habit of making a noise at their hinder quarters, as, for instance, the turtle-dove; and they make a violent movement of their tails at the same time that they produce this peculiar sound.

Printed in Great Britain
by Amazon